European Geophysical Society Series on Hydrological Sciences, 1

Advances in Theoretical Hydrology

A Tribute to James Dooge

European Geophysical Society Series on Hydrological Sciences, 1

Advances in Theoretical Hydrology

A Tribute to James Dooge

Edited by

J. Philip O'Kane

Department of Civil and Environmental Engineering,
University College Cork, Cork, Ireland

ELSEVIER
Amsterdam – London – New York – Tokyo 1992

ELSEVIER SCIENCE PUBLISHERS
Sara Burgerhartstraat 25
P.O. Box 211, 1000 AE Amsterdam, The Netherlands

Library of Congress Cataloging-in-Publication Data

Advances in theoretical hydrology : a tribute to Jim Dooge / edited by
J. Philip O'Kane.
p. cm. -- (European Geophysical Society series on
hydrological sciences ; 1)
Includes bibliographical references and index.
ISBN 0-444-89831-X (acid-free)
1. Hydrology. 2. Dooge, James. I. Dooge, James. II. O'Kane, J.
P., 1943- . III. Series.
GB662.4.A38 1992
551.48--dc20 92-35090
CIP

ISBN: 0 444 89831 X

This book is printed on acid-free paper.

Printed in The Netherlands.

Professor James C.I. Dooge and his late wife Ronnie

Preface

It has been my pleasure and privilege to assemble this collection of sixteen essays and papers as a tribute to the distinguished Irish hydrologist James Dooge on the occasion of his 70th birthday. His former students, his colleagues and friends in fourteen countries, replied to my invitation to contribute to the Festschrift by providing a delightfully varied response. It includes those favourite topics of his, flow in open channels and unsaturated soil, and also his major interest of recent years, large-scale hydrology and global change.

The ten scientific papers fall naturally in two sections: hydrological processes and large-scale hydrology. The six historical, reflective and philosophical essays, on the past and future of the hydrological sciences, form the third and final section of the book.

I have also included a brief curriculum vitae and list of publications for those younger readers, who would like to study James Dooge's contribution to the different areas of hydrology, to the promotion of international scientific co-operation, and to the recognition of hydrology as a fundamental earth science.

The extent of his scientific influence is evident in these pages, but the contributors also wished to pay tribute to his personal inspiration and human qualities.

Manabe insists that James Dooge is one of the few visionaries, who recognised long ago the importance of close collaboration between hydrologists and atmospheric and oceanic scientists for the study of climate change. Becker has reminded us of the help that James Dooge always very willingly gave to researchers in various parts of eastern Europe to make their research results better known in the West. McBean underlines the major role that James Dooge played in initiating and formulating the land-surface component of the World Climate Research Programme, and in particular the Hydrologic-Atmospheric Pilot Experiments (HAPEX) and the Global Energy and Water Cycle Experiment (GEWEX) which are amongst its most important parts. Wang reminds us the James Dooge has always been greatly respected by his Chinese colleagues for his ingenuity, enthusiasm, wide interests, hard-work and good humour.

It is clear from all the tributes, that his friends and colleagues have always appreciated, not only his scientific stature, but also his personal courtesy, generosity in sharing insights, and openness to new ideas. I think we should also note that his remarkable scientific career has been achieved at the same time as a very successful political career in Ireland.

In many ways this book is also a tribute to Ronnie Dooge whose wide range of interests and support for hydrology was appreciated by all of us. I hope she would have approved of this book.

Tá a scéal fhéin sa leabhair seo freisin. Ar dheis Dé go raibh a anam.

Finally, as a former student and colleague of James Dooge and as the holder of a University chair which he once occupied, I join his numerous friends in the world of hydrology and congratulate him on his 70th birthday.

Ad Multos Annos.

J. PHILIP O'KANE
Department of Civil and Environmental Engineering
University College Cork

July 1992

James Dooge — Curriculum vitae

1984–	Research Consultant, Centre for Water Resources Research, University College, Dublin.
1970–1984	Professor of Civil Engineering, University College, Dublin.
1958–1970	Professor of Civil Engineering, University College, Cork.
1946–1958	Design Engineer, Civil Works Department of Electricity Supply Board, Dublin.
1943–1946	Junior Engineer on river improvement works, Arterial Drainage Section of Office of Public Works.

Member of the Bureau of the Organising Committee for the First World Climate Conference (February, 1970).
Chairman of the Scientific Advisory Committee for the World Climate Impact Studies Programme of the U.N. Environmental Programme since its formation in 1980.
Member of Ad Hoc Group of Experts established by the Secretary General of the U.N. to advise him on the establishment of an International Decade for Natural Disaster Reduction.
Chairman of the Organising Committee for the Second World Climate Conference (October/November 1990).
Chairman of the Evaluation Panel on the Environmental Research Programme of the European Community (1987–1988).
Special Consultant to the European Commission on the relationship between research outputs and policy formulation (1989–1990).

B.E., B.Sc. (NUI, 1942), M.E. (NUI, 1952), M.Sc. (Iowa, 1956).

Doctor of Agricultural Sciences (h.c. Wageningen, 1978). Doctor of Technology (h.c. Lund, 1980). Doctor of Science (h.c. Birmingham, 1985). Doctor of Science (h.c. Dublin, 1988).

Honorary Fellow of the Institution of Engineers of Ireland (President, 1968-69; Kettle Premium and Plaque, 1948; Mullins Medal, 1951; Mullins Medal, 1962; Kettle Plaque, 1985).
Member of the Royal Irish Academy (Secretary, 1978–1981; President, 1987–1990).
Honorary Member of the Irish Meteorological Society.
Honorary Life Member of the Royal Dublin Society.

Fellow of the Institution of Civil Engineers (London).
Fellow of the American Society of Civil Engineers.
Fellow of the American Geophysical Union (Horton Award, 1959; Bowie Medal, 1986).
Foreign Member of the Polish Academy of Sciences.
Foreign Member of the Accademia Patavina (Padua).
Fellow of the Collegium Ramazzini (Modena).
Honorary Member, Hungarian Hydrological Society.

Honorary Professor of Ho Hai University (East China Technical University of Water Resources).
Awarded the 1990 Leonardo da Vinci Medal of the European Society for Engineering Education.

President, International Commission on Water Resources, 1971–1975.
President, International Association for Hydrological Sciences, 1975–1979.
Bureau Member of the International Union of Geodesy and Geophysics, 1979–1987.
General Committee of the International Council of Scientific Unions, 1980–1986 and 1988–1992 (Secretary General, 1980–1982; President-Elect, 1990–).

Elected a member of the Irish Senate in 1961.
Deputy Chairman, 1965–1973; Chairman, 1973–1977.
Minister for Foreign Affairs, 1981–1982.
Government Leader in the Senate, 1983–1987.

Chairman of the ad hoc Committee of personal representatives of EEC Heads of Government whose report of March 1985 on Institutional Reform and European Union was the basis of the inter-governmental negotiations leading to the Single European Act.

James Dooge — List of publications

Publications and discussions (sole author)

1947 Discussion of *"Tests on models of the River Erne Power Scheme" by G. FitzGibbon and J.A. O'Riordan*. Transactions, Institution of Civil Engineers of Ireland, Vol. 73, p. 237.

1948 *The Use of Models as an Aid to Engineering Design*. This paper (24 pages) was awarded the Laurence J. Kettle Premium and Plaque of the Institution of Civil Engineers of Ireland.

1949 Discussion of *"Run-off of the River Liffey and the Lower River Erne" by J.A. O'Riordan*. Transactions, Institution of Civil Engineers of Ireland, Vol. 75.

1952 *Unsteady Flow in Open Channels*. Thesis, for Degree of Master of Engineering. National University of Ireland, 153 pp.

1954 Discussion of *"Recent developments in the water supply to the City of Dublin" by G.A. Cremins*. Transactions, Institution of Civil Engineers of Ireland, Vol. 80, pp. 250–251.

1954 Discussion of *"Analysis of 35-year automatic recordings of rainfall at Cork" by E.C. Dillon*. Transactions, Institution of Civil Engineers of Ireland, Vol. 80, pp. 271–273.

1955 Discussion of *"Hydraulic drop as a function of velocity distribution" by C.G. Edson*. Civil Engineering (February 1955), p. 100, New York.

1955 Correspondence on *"The employment of unit hydrographs to determine the flows of Irish arterial drainage channels" by J.J. O'Kelly*. Proceedings, Institution of Civil Engineers, Vol. 4(iii), pp. 436–442, London.

1956 *Synthetic Unit Hydrographs based on Triangular Inflow*. Thesis for Master of Science Degree at the University of Iowa, 104 pp.

1957 *Stability of Jets from Ski-Jump Sluices: Model Tests for Carrigadrohid Dam*. International Association for Hydraulic Research, Transactions of the Lisbon Congress, Vol. 11, D.5, pp. 1–10.

1957 *The Rational Method for Estimating Flood Peaks — Irish Contributions to the Technique*. Presented at the Dublin Meeting of the British Association for Advancement of Science. Reprinted in Engineering, September 6, pp. 311–313, and September 20, pp. 374–377.

1957 Discussion of *"The construction of the Sasumua Dam in Kenya, East Africa"*. Transactions, Institution of Civil Engineers of Ireland, Vol. 83, pp. 136–139.

1958 Discussion of letter by R.W. Powell on the Manning formula. Civil Engineering (June 1958), p. 441, New York.

1958 Discussion of *"Determining run-off from rainfall" by J.E. Nash*. Proceedings Institution of Civil Engineers, Vol. 11, pp. 511–514, London.

1959 *Operations Research*. Administration. Vol. 6, No. 4, pp. 346–361.

1959 *A General Theory of the Unit Hydrograph*. Journal of Geophysical Research, Vol. 64, No. 2, pp. 241–256. Received the Robert E. Horton Award for 1959 from the American Geophysical Union, "for a paper of outstanding excellence in the field of hydrology".

1959 *Un Bilan Hydrologique au XVIIe Siecle*. La Houille Blanche, November 1959, pp. 799–807.

1959 Discussion of *"Discharge from heavy rainfall" by E.E. Dawson*. Proceedings Institution of Civil Engineers, Vol. 12, pp. 456–457, London.

1959 Discussion of *"Rural Water Supplies" by T.J. O'Riordan*. Transactions, Institution of Civil Engineers of Ireland, Vol. 85, pp. 147–149.

1960 *The Routing of Groundwater Recharge Through Typical Elements of Linear Storage*. Commission on Subterranean Water, International Association for Scientific Hydrology, Proceedings of Helsinki General Assembly, IAHS Publication No. 52, pp. 286–300.

1960 *What is Administrative Efficiency? An Engineer's View*. Administration. Vol. 8, No. 4, pp. 251–258.

1961 Discussion of *"A unit hydrograph study, with particular reference to British catchments" by J.E. Nash*. Proceedings Institution of Civil Engineers, Vol. 20, pp. 467–470, London.

1961 Discussion of *"Education and training of professional engineers" by M.A. Hogan*. Transactions, Institution of Civil Engineers of Ireland, Vol. 87, p. 222.

1962 *An Indirect Method of Estimating Low Flows*. Institution of Civil Engineers of Ireland, Transactions, Vol. 87, pp. 125–151. Awarded a Mullins Silver Medal.

1963 Discussion of *"Some aspects of the hydrology of the Thames basin" by F.M. Andrews*. Proceedings Institution of Civil Engineers, Vol. 24, pp. 252–254.

1965 *Analysis of Linear Systems by Means of Laguerre Functions*. Society for Industrial and Applied Mathematics, Journal on Control, Vol. 2, No. 3, pp. 396–408.

1966 *The Education of the Technologist.* University Review. Vol. 3, No. 7, pp. 39–45.

1967 *The Hydrologic Cycle as a Closed System.* Special Invitation Lecture at Fort Collins International Symposium, Proceedings, Vol. 2, pp. 98–113. Reprinted in IAHS Bulletin, Vol. XIII, No. 1, pp. 58–66, 1968.

1967 *A New Approach to Non-Linear Problems in Surface Water Hydrology.* Commission on Surface Waters, International Association for Scientific Hydrology, Proceedings of General Assembly of Berne, IAHS Publication No. 76, pp. 409–413.

1968 *Presidential Address.* Institution of Civil Engineers of Ireland, Transactions, Vol. 95, No. 1, pp. 3–9.

1968 Discussion of *"An investigation of the applicability of linear programming to the design of flexible pavements" by T.M. McCann.* Transactions, Institution of Engineers of Ireland, Vol. 94, pp. 165–166.

1970 *Peat Silt as a Sediment.* Salmon Research Trust of Ireland, Annual Report No. XIV, pp. 67–75.

1970 Discussion of *"Flood estimation for ungauged catchments" by M.A. Lynn.* Transactions, Institution of Engineers of Ireland, Vol. 96, pp. 169–170.

1972 *Mathematical Models of Hydrologic Systems.* Proceedings of International Symposium on Modelling Techniques in Water Resource Systems, Vol. 1, pp. 171–189. Department of the Environment, Ottawa, Ont., Canada.

1972 *Conceptual Models of Surface Runoff.* Atti del Convegno Internazionale Sul Tema, Piene: Loro Previsione e Difesa Del Suolo, Quaderno N. 169, pp. 179–207, Academic Nazionale dei Lincei, Rome.

1973 *Linear Theory of Hydrologic Systems* U.S. Agricultural Research Service, Technical Bulletin, No. 1468 (October), 327 pp., Washington, D.C.

1974 *Modele Matematyczne Systemow Hydrologicznych.* In: J. Kindler (Editor), Matematyczne Modele Siewni. Polish Academy of Sciences, Wroclaw, Chapter 2, pp. 53–79.

1974 *Wybor Modelu Deterministycznego.* In: J. Kindler (Editor), Matematydzne Modele Zlewni. Polish Academy of Sciences, Wroclaw, Chapter 3, pp. 80–110.

1975 *The Water Balance of Bogs and Fens.* Proceedings of the Minsk Symposium on the Hydrology of Marsh-Ridden Areas. IHD Studies and Reports on Hydrology, No. 19, pp. 233–272, UNESCO, Paris.

1975 Scientific Editor of *Application of Mathematical Models in Hydrology and Water Resource Systems*, Pre-printed papers for International Symposium at Bratislava, IAHS Publication No. 115, 284 pp., International Association of Hydrological Sciences, September.

1975 *Report on the activities of the International Commission on Water Resources Relations and Systems for the period 1971–1975*. Hydrological Sciences Bulletin, Vol. 20, No. 4, pp. 433–435, International Association of Hydrological Sciences, December.

1976 Scientific Editor of Late papers, general reports, and reports on the workshops, Bratislava Symposium, Hydrological Sciences Bulletin, Vol. 21, No. 1, pp. 3–229, International Association of Hydrological Sciences, March.

1976 Scientific Editor of *Mathematical Models in Geophysics*, Proceedings of Moscow Symposium, IAHS Publication No. 116, 215 pp., International Association of Hydrological Sciences.

1977 *Problems and Methods of Rainfall-Runoff Modelling*. In: T.A. Ciriani, U. Maione and J.R. Wallis (Editors), Mathematical Models for Surface Water Hydrology. Proceedings of Workshop held at IMH Center, Piza, Italy. Wiley, New York, N.Y., pp. 71–108,

1977 *The Contribution of Engineers to the Economic Development of Ireland*, Irish Engineers, Vol. 30, No. 9 (October), pp. 19–21.

1978 *Mathematical Models in Surface Hydrology*, IBM Technical Report 6513-3539 (August), IBM Scientific Centre, Pisa.

1978 *Linear Algebra and Unit Hydrograph Stability*, International Symposium on Hydrology and Water Resources, Simon Bolivar University, Venezuela, March 1978.

1979 *Deterministic Input–Output Models*. In: E.H. Lloyd, T. O'Donnell and J.C. Wilkinson (Editors), Mathematics of Hydrology and Water Resources, Proceedings of Lancaster Symposium on Mathematics of Hydrology and Water Resources. Academic Press, London, pp. 1–37.

1979 *Suiwen Sitong Te Siensing Liteng*, Chinese translation of USDA Monograph *Linear Theory of Hydrologic Systems* (1973).

1979 *Alternative Approaches to Flow Problems*. General Lecture to Cagliari Congress of International Association for Hydraulic Research, September 1979. Proceedings of XVIII IAHR Congress, Vol. 6, pp. 28–55.

1979 *Deterministic Methods in Systems Hydrology*. East China College of Hydraulic Engineering, 250 pp. (October 1979).

1980 *Water Research and Human Problems*. Bulletin de COWAR, 1980, pp. 44–54, International Council of Scientific Unions, Paris.

1980 *Flood Routing in Channels*. Department of Civil Engineering, University College, Dublin Postgraduate Lecture Notes, 259 pp. (August 1980).

1981 *Model Structure and Classification*. In: A.J. Askew, F. Greco and J. Kindler (Editors), Logistics and Benefits of Using Mathematical Models of Hydrologic and Water Resource Systems. Pergamon, Oxford, pp. 1–25.

1982 *Parametrization of Hydrologic Processes*. In: P.S. Eagleson (Editor), Land Surface Processes in Atmospheric General Circulation Models. Cambridge University Press, Cambridge, pp. 243–288.

1983 *On the Study of Water*. Jubilee Lecture to Exeter General Assembly of IAHS (July 1982), Hydrological Sciences Journal, Vol. 28, No. 1 (March 1983), pp. 23–48.

1983 *Unit Hydrograph Concepts and Methods*. Department of Civil Engineering, University College, Dublin (June, 1983), Postgraduate Lecture Notes, 208 pp.

1983 *Theory of Flood Routing*. Postgraduate Course at Agricultural University of Wageningen (July, 1983). Published as Chapter 3 in: D.A. Kraijenhoff van de Leur and J.R. Moll (Editors), River Flow Modelling and Forecasting. Reidel, Dordrecht, pp. 39–65, 1986.

1983 *Waters of the Earth*. Union Lecture delivered at Hamburg, General Assembly of IUGG (August 1983). Published in Hydrological Sciences Journal, Vol. 29, No. 2, pp. 149–176 (June 1984) and also published in Geo. Journal 8.4, pp. 325–340 (1984)

1984 *Water and the Third World*. In: Proceeding of World Food Day Colloquium. GORTA (Freedom from Hunger Council of Ireland), Dublin, pp. 13–16.

1984 *Hydrological Inputs to Global Climate Models*. UNU Workshop on Forests, Climate and Hydrology, Oxford, March 1984. Published as Chapter 7 in: E.R.C. Reynolds and F.B. Thompson (Editors), Forests, Climate and Hydrology. United Nations University, Tokyo, pp. 129–161, 1988.

1985 *Historical Development of Open Channel Flow*. IAHR 50th Anniversary Symposium, Berlin, April 1985. In: G. Garbrecht (Editor), Hydraulics and Hydraulic Research: A Historical Review. Balkema, Rotterdam, pp. 205–230, 1987.

1985 *Manning and Mulvany. River Improvement in 19th Century Ireland*. IAHR 50th Anniversary Symposium, Berlin, April 1985. In: G. Garbrecht (Editor), Hydraulics and Hydraulic Research: A Historical Review. Balkema, Rotterdam, pp. 173–183, 1987.

1985 *Technology Transfer in Water Engineering*. FEANI/IEI International Seminar on Case Studies in Technological Transfer, Dublin, September 1985. Awarded a Mullins Silver Medal.

1985 *Looking for Hydrologic Laws.* Keynote Address to U.S./China Bilateral Symposium on the Analysis of Extraordinary Flood Events. Published in Water Resources Research, Vol. 22, No. 9, pp. 465–585, August 1986.

1985 *Hydrologic Models of Land Surface Processes.* ISLSCP Conference, Rome, December 1985. Published in ESA SP-248, pp. 3–16, European Space Agency, May 1986.

1986 *Scale Problems in Hydrology,* Kisiel Memorial Lecture, February, 1986. University of Arizona, Tucson, Ariz., 1989, 63 pp.

1986 *Climate Change and Water Resources,* Special Lecture to WMO Executive Council, June 1986. WMO Publication No. 675, pp. 37–62, Geneva, 1987.

1986 *Evaluation of Irish Aid Effort.* In: Long Term Development, Proceedings of World Food Day Colloquium 1986. GORTA (Freedom from Hunger Council of Ireland), Dublin, pp. 9–22

1986 *Effects of CO_2 Increases on Hydrology and Water Resources,* E.E.C. Symposium on CO_2 and other Cases. Brussels, November 1986. Published in: R. Fantechi and A. Ghazi (Editors), Carbon Dioxide and other Greenhouse Gases: Climatic and Associated Impacts Kluwer, Dordrecht, pp. 204–213, 1989.

1987 Keynote Lecture to IAHS/IAHR International Symposium on Water for the Future, Rome, April 1987. Published by IAHS as *Hydrology in Perspective,* Hydrological Sciences Journal, 33(1) pp. 61–85. Published by IAHR as *Hydrology — Past and Present,* Journal of Hydraulic Research, Vol. 26(1), pp. 5–26.

1989 *The flow of Irish rivers.* In: M. Steer (Editor), Irish Rivers: Biology and Management. Royal Irish Academy, pp. 5–26, 1991.

1989 *The role of the hydrological cycle in climate.* In: A. Berger, S. Schneider and J.C.I. Duplessy (Editors), Climate and Geosciences. Kluwer, Dordrecht, 1989.

1989 *The interaction of the IGBP with ICSU bodies and with other international programmes.* In: Report of the First Meeting of the Scientific Advisory Council of IGBP. IGBP Publication 7:1, pp. 97–101, Stockholm.

1989 *Water and the Third World.* Sixth O'Sullivan Lecture, 24th November 1988. Imperial College of Science and Technology, London, 21 pp.

1989 *The Manning formula in context.* In: B.C. Yen (Editor), Channel Flow and Catchment Runoff. International Conference for Centennial of Manning's Formula and Kuichling's Rational formula. University of Virginia, Charlottesville, May 1989, pp. 849–902.

1989 *Robert Manning 1816–1895*. Institution of Engineers of Ireland, December 1989, 79 pp.

1990 *Da Vinci, Engineering and Europe*. In: V.J. McCabe (Editor), Design in Engineering Education. Proceeding of SEFI Annual Conference. Société Européenne pour la Formation des Ingénieurs. Brussels, pp. 509–516.

1990 *Ireland and its infrastructure*. In: J.W. de Courcy (Editor), Proceedings of Conference on Aspects of the Irish Infrastructure — Present Condition and Future Needs. Department of Civil Engineering, University College, Dublin, pp. 347–354.

1990 *The World Climate Impact Studies Programme*. In: J. Jager and H. Ferguson (Editors), Climate Change: Science, Impacts and Policy. Proceeding of Second World Climate Conference. Cambridge University Press, pp. 169–175, 1991.

1990 *Hydrologic models and climate change*. Keynote Address to AGU Chapman Conference on Hydrological Aspects of Global Climate Change, Lake Chelan, Washington. Journal of Geophysical Research, Vol. 97(D3): 2677–2686. American Geophysical Union, February 1992.

1990 *De Marchi's legacy to hydrology*. Celebration of the centennial of the birth of Giulio de Marchi (1890–1990). Politecnico di Milano, October 1990 (in press).

1991 *Water and development 1981–1990*. Water and Environmental Section, Institution of Engineers of Ireland, January 1991.

1991 *Sensitivity of runoff to climate change: a Hortonian approach*. Robert E. Horton Memorial Lecture. Annual Meeting of American Meteorological Society, New Orleans, La., January 1991 (to be published in Bulletin of AMS).

1991 Introductory remarks to EURISY Symposium on *The Earth's Environment: An Assessment from Space*, Venice, April 1991, Proceedings Volume, pp. 13–14. European Space Agency.

1991 *Climate change, actual evaporation and catchment runoff*. Contribution to Scientific Session of the 25th Anniversary of the International Course for Hydrologists. International Institute for Hydraulic and Environmental Engineering, Delft.

1992 *Climate change and natural disasters: where are the links?* Proceedings of Symposium on *The World at Risk: Natural Hazards and Climate Change*, M.I.T. Center for Global Change, January, 1992 (to be published by American Institute of Physics).

1992 *Scale problems in surface infiltration and exfiltration*. George Kovacs Colloquium on Space and Time Scale Variability and Interdependencies in Various Hydrologic Processes. UNESCO, July 1992.

Publications and discussions (joint author)

1951 Dooge, J.C.I. and Murphy, A.M., *The Hydraulic Fish Lift at Leixlip.* Institution of Civil Engineers of Ireland Trans., Vol. 22, pp. 61–68. Awarded a Mullins Silver Medal.

1955 Delleur, J.W., Dooge, J.C.I. and Ghent, K.W., *Influence of Slope and Roughness on the Free Overfall*, Reproduced in: Seven Exploratory Studies in Hydraulics, Proceedings of American Society of Civil Engineers, Paper 1038, p. 2 and pp. 30–35.

1967 Dooge, J.C.I. and Harley, B.M., *Linear Routing in Uniform Open Channels*, Fort Collins International Symposium Proceedings, Vol. 2, pp. 98–113.

1968 Harley, B.M. and Dooge, J.C.I., *Problems in Simulating and Evaluating Various Methods of Flood Routing Using a Small Digital Computer*, Tucson International Symposium on the Use of Analog and Digital Computers in Hydrology, IAHS Publication No. 8, pp. 417–425.

1971 Dooge, J.C.I., Kalinin, G.P. and Dawdy, D.R., *Report of Committee on Mathematical Models in Hydrology*, Bulletin of IAHS Vol. XVI, No. 4, pp. 9–55 (December 1971).

1972 Dooge, J.C.I., Costin, A.B. and Finkel, H.J., *Man's Influence on the Hydrological Cycle.* Irrigation and Drainage Paper No. 17, UN Food and Agricultural Organisation, Rome.

1974 Dooge, J.C.I., Costin, A.B. and Finkel, H.J., *Manniskin i Vattnets Kletsiopp.* Swedish translation of FAO Paper 17, Lund, Sweden.

1975 Keane, R. and Dooge, J.C.I., *The Effects of Initial Moisture Content on Infiltration into Peat*, Proceedings of the Minsk Symposium on the Hydrology of Marsh-Ridden Areas, IHD Studies and Reports in Hydrology, No. 19, pp. 274–280, UNESCO, Paris.

1975 Dooge, J.C.I. and Keane, R., *Mathematical Simulation of Runoff from Small Plots of Undrained and Drained Peat at Glenamoy*, Proceedings of the Minsk Symposium on the Hydrology of Marsh-Ridden Areas. IHD Studies and Reports in Hydrology, No. 19, pp. 369–377, UNESCO, Paris.

1977 Dooge, J.C.I. and O'Kane, J.P.J., *PICOMO: A programme for the identification of conceptual models.* In: T.A. Ciriani, U. Maione, and J.R. Wallis (Editors), Mathematical Models for Surface Water Hydrology. Proceedings of Workshop held at IBM Center, Piza, Italy. Wiley, New York.

1978 Dooge, J.C.I. and Garvey, G.J., *The Use of Meixner Functions in the Identification of Heavily Damped Linear Systems*, Proceedings of the Royal Irish Academy, Vol. 78(A), No. 18, pp. 157–179.

1978 Dooge, J.C.I. and O'Kane, J.P.J. (Editors), *Water Balance of Europe.* Key Reports of 1976 Varna Workshop. Technical Documents in Hydrology, UNESCO, Paris.

1979 Dooge, J.C.I. and Dowley, A., *Rainfall and Runoff.* In: Road Drainage in Ireland. An Foras Forbartha (National Planning Institute), Dublin, pp. 15–42.

1981 Dooge, J.C.I. and Lynch, P.J., *Staff Requirements for Engineering Teaching.* Second Conference on Manpower Development. National Board for Science and Technology, Dublin.

1981 Dooge,j.C.I., Strupczewski, W.G. and Napiorkowski, J.J., *The Use of Volterra Series in Modelling of Flow in an Open Channel.* International Conference on Numerical Modelling of Rivers, Channels and Overland Flow for Water Resources and Environmental Applications, Bratislava, May 1981.

1982 Napiorkowski, J.J., Strupczewski, W.G. and Dooge, J.C.I., *Lumped Non-Linear Flood Routing Model and its Simplification to the Muskingum Model.* International Symposium on Rainfall–Runoff Modelling Mississippi State University, May 1981. Published in: V.P. Singh (Editor), Rainfall–Runoff Relationships. Water Resources Publications, Fort Collins, Colo.

1983 Dooge, J.C.I. and Maczuga, T., *Computer Comparison of One-Dimensional Flood Routing.* Bulletin of Polish Academy of Sciences, Vol. XXX, Nos. 11/12, pp. 21–26.

1983 Dooge, J.C.I., Kundzewicz, Z.W. and Napiorkowski, J.J., *On the Backwater Effect in Linear Diffusion Routing.* Hydrological Sciences Journal, Vol. 28, No. 3, pp. 391–402.

1984 Dooge, J.C.I. and Baker, F.W.G., *Combining the Strength of Governmental and non-Governmental Approaches.* Chapter 29 in: F. di Castri, F.W.G. Baker and M. Hadley (Editors), Ecology in Practice. Tycooley Press, Dublin, pp. 334–342.

1984 Bruen, M. and Dooge, J.C.I., *An Efficient and Robust Method for Estimating Unit Hydrograph Parameters.* Journal of Hydrology, Vol. 70, pp. 1–24.

1984 Dooge, J.C.I. and Napiorkowski, J.J., *Effect of downstream control in flood routing*. Acta Geophysica Polonica, Vol. 32, pp. 363–373.

1985 Kundzewicz, Z.W. and Dooge, J.C.I., *Unified Structural Approach to Linear Flood Routing*. Advances in Water Resources, Vol. 8, pp. 37–42.

1985 De Buitléar, E. (Editor) with contributions by Dooge, J.C.I., Goodwillie, R., Heuff, H. and Whelan, K., *Irish Rivers*. Wild Ireland Library, Criterion Press, Dublin.

1986 Dooge, J.C.I., O'Kane, J.P.J. and Brandyk, T., *Soil moisture in general circulation models*. Acta Geophysica Polonica, Vol. 34, No. 4, pp. 337–356.

1987 Dooge, J.C.I. and Napiorkowski, J.J., *Applicability of diffusion analogy in flood routing*. Acta Geophysica Polonica, Vol. 35, No. 1, pp. 65–75.

1987 Dooge, J.C.I. and Napiorkowski, J.J., *The effect of the downstream boundary condition in the linearized St. Venant equations*. Quarterly Journal of Mechanics and Applied Mathematics, Vol. 40(2), pp. 245–256.

1987 Dooge, J.C.I., Napiorkowski, J.J. and Strupczewski, W.G., *The linear downstream response of a generalised uniform channel*. Acta Geophysica Polonica, Vol. 35, No. 3.

1987 Dooge, J.C.I., Napiorkowski, J.J. and Strupczewski, W.G., *Properties of the generalised linear downstream channel response*. Acta Geophysica Polonica, Vol. 35, No. 4.

1988 Napiorkowski, J.J. and Dooge, J.C.I., *Analytical solution of channel flow model with downstream control*. Hydrological Sciences Journal, Vol. 33(3), pp. 269–287.

1988 Romanowicz, R.J., Dooge, J.C.I. and Kundzewicz, Z.W., *Moments and cumulants of linearised St. Venant equation*. Advances in Water Resources, Vol. 11, pp. 92–100.

1989 Kundzewicz, Z.W. and Dooge, J.C.I., *Attenuation and phase shift in linear flood routing*. Hydrological Sciences Journal, Vol. 34(1), pp. 121–139.

1989 Turner, J., Bree, T. and Dooge, J.C.I., *Deriving the unit hydrograph by root selection*. Journal of Hydrology, Vol. 110, pp. 137–152.

1989 Strupczewski, W.G., Napiorkowski, J.J. and Dooge, J.C.I., *The distributed Muskingum model*. Journal of Hydrology, Vol. 111, pp. 235–257.

1989 Dooge, J.C.I. and Bruen, M., *Unit hydrograph stability and linear algebra*. Journal of Hydrology, Vol. 111, pp. 377–390.

1990 Kuhnel, V., Dooge, J.C.I., Sander, G.C. and O'Kane, J.P.J., *Duration of atmosphere controlled and of soil controlled phases of infiltration for constant rainfall at a soil surface.* Annales Geophysicae, Vol. 8(1), pp. 11–20.

1990 Kuhnel, V., Dooge, J.C.I., Sander, G.C. and O'Kane, J.P.J., *Duration of atmosphere controlled and soil controlled phases of evaporation for constant potential evaporation at a soil surface.* Annales Geophysicae, Vol. 8(1): pp. 21–28.

1991 Chuta, P. and Dooge, J.C.I., *Deterministic derivation of the geomorphic unit hydrograph.* Journal of Hydrology.

1991 Agsorn, S. and Dooge, J.C.I., *Numerical experiments on the monoclinal rising wave.* Journal of Hydrology, Vol. 124: pp. 293–306.

1991 Zhao Di Hua and Dooge, J.C.I., *A simple conceptual model of subsurface conditions.* In: Water for Life. Proceeding of Silver Jubilee Seminar of International Centre of Hydrology, University of Padua, pp. 34–46.

1991 Kuhnel, V., Dooge, J.C.I., O'Kane, J.P.J. and Romanowicz, R.J., *Partial analysis applied to scale problems in surface moisture fluxes.* Surveys in Geophysics, Vol. 12: pp. 221–247.

1992 Dooge, J.C.I., Goodman, G.T., La Riviere, J.W.M., Marton-Lefevre, J., O'Riordan, T. and Praderie, F. (Editors), *An Agenda of Science for Environment and Development into the 21st Century.* Cambridge University Press, 352 pp.

1992 Dooge, J.C.I., Perumal, M. and Wang, Q.J., *Step function response of a Muskingum reach.* Technical Note, ASCE Journal of Irrigation and Drainage Engineering (in press).

1992 Bruen, M. and Dooge, J.C.I., *Unit hydrograph estimation with multiple events and prior information. I: Theory and a computer program* (accepted by Journal of Hydrological Sciences).

1992 Bruen, M. and Dooge, J.C.I., *Unit hydrograph estimation with multiple events and prior information. II: Evaluation of the method.* (accepted by Journal of Hydrologic Sciences).

1992 Wang, Q.J. and Dooge, J.C.I., *Limiting cases of land surface fluxes.* XVI General Assembly of European Geophysical Society, Wiesbaden, April 1991 (submitted to Journal of Hydrology).

1992 Dooge, J.C.I. and Wang, Q.J., Comment on *"An investigation of the relationship between ponded and constant flux rainfall infiltration" by A. Poulavassilis, P. Kerkides, S. Elnaloglou and I. Argyrokastritis (Water Resources Research 27(7): 1403–1409, 1991)* (submitted to Water Resources Research).

1992 Dooge, J.C.I. and Napiorkowski, J.J., Discussion of *"Study of open channel flow as controlled process" by Y.A. Ermolin*. ASCE Journal of Hydraulic Engineering, Vol. 118(1), pp. 59–72.

List of contributors

M.B. ABBOTT	*International Institute for Hydraulic and Environmental Engineering, Oude Delft 95, 2601 DA Delft, The Netherlands*
A. BECKER	*BAHC Office, Core Projects Office, Institut für Meteorologie, Freie Universität Berlin, Dietrich-Schäfer-Weg 10, D-1000 Berlin 41, Germany*
K. BRYAN	*GFDL/NOAA, Princeton University, Forrestal Campus, P.O. Box 308, Princeton, NJ 08542, U.S.A.*
P. S.P. COWPERTWAIT	*Department of Civil Engineering, University of Newcastle upon Tyne, Cassie Building, Newcastle upon Tyne NE1 7RU, England*
L. DÜMENIL	*Max Planck Institute for Meteorology, Bundesstraße 55, D-2000 Hamburg 13, Germany*
P. S. EAGLESON	*Ralph M. Parsons Laboratory, Department of Civil Engineering Building 48-335, Massachusetts Institute of Technology, Cambridge, MA 02139, U.S.A.*
L.T. GOTTSCHALK	*University of Oslo, Institute of Geophysics, P.O. Box 1022, Blindern, 0315 Oslo 3, Norway*
L.S. KUCHMENT	*Laboratory of Modelling Hydrological Processes, Water Problems Institute of USSR Academy of Sciences, Moscow 103064*
G.A. McBEAN	*Atmospheric Science Programme and Department of Oceanography, University of British Columbia, 194 West Mall, Vancouver BC, V6T 1Z2, Canada*
S. MANABE	*GFDL/NOAA, Princeton University, Forrestal Campus, P.O. Box 308, Princeton, NJ 08542, U.S.A.*
J. NAPIÓRKOWSKI	*Humboldt Fellow, Ruhr Universität, Institut für Hydrologie und Wasser, P.O Box 102146, D-4630 Bochum 1, Germany* On leave from: *Institute of Geophysics, Polish Academy of Sciences, Warsaw, Poland*
J.E. NASH	*Department of Engineering Hydrology, University College, Galway, Ireland*

P.E. O'CONNELL — *Department of Civil Engineering, University of Newcastle upon Tyne, Cassie Building, Newcastle upon Tyne NE1 7RU, England*

J.P. O'KANE — *Department of Civil and Environmental Engineering, University College, Cork, Ireland*

J.R. PHILIP — *Centre for Environmental Mechanics, CSIRO, G.P.O. Box 821, Canberra ACT 2601, Australia*

M.J. SPELMAN — *GFDL/NOAA, Princeton University, Forrestal Campus, P.O. Box 308, Princeton, NJ 08542, U.S.A.*

R.J. STOUFFER — *GFDL/NOAA, Princeton University, Forrestal Campus, P.O. Box 308, Princeton, NJ 08542, U.S.A.*

M. SUGAWARA — *6-13-30 Minami-Karasuyama, Setagaya-ku, Tokyo 157, Japan*

E. TODINI — *Institute of Hydraulic Construction, Faculty of Engineering, University of Bologna, Viale del Risorgimento, 2, 40136 Bologna, Italy*

G. VACHAUD — *Institut de Mécanique de Grenoble, Domaine Universitaire, St. Martin d'Heres, BP 53 X, 38041 Grenoble Cédex, France*

M. VAUCLIN — *Institut de Mécanique de Grenoble, Domaine Universitaire, St. Martin d'Heres, BP 53 X, 38041 Grenoble Cédex, France*

Q.J. WANG — *Centre for Water Resources Research, Department of Civil Engineering, University College, Earlsfort Terrace, Dublin 2, Ireland*

Contents

Preface VII
James Dooge — Curriculum vitae IX
James Dooge — List of publications XI
List of contributors XXIII

PART A. HYDROLOGICAL PROCESSES

Chapter 1. Linear theory of open channel flow
by Jarosław J. Napiórkowski 3
Abstract 3
1. Introduction 3
2. Linearization of the St. Venant equations 4
3. Solution for finite channel reach 6
4. The generalized linear channel response 10
   4.1. Cumulants of generalized channel response 10
   4.2. Amplitude and phase characteristics 12
5. Simplified forms of St. Venant equations 13
6. Conclusions 14
References 14

Chapter 2. Analytical and numerical modelling of unsaturated flow
by Q.J. Wang 17
Abstract 17
1. Introduction 17
2. Physics of soil moisture flow 18
3. Reduced form of Richards' equation 19
4. Closed form solutions for constant D and linear K^* model 20
5. Effect of assumptions on D and K^* functions 23
6. A proposed hybrid solution scheme for future research 24
References 25

Chapter 3. The hydrology of milled peat production
by J. Philip O'Kane 27
Abstract 27
1. Introduction 27
2. The peat–water system 28

2.1. Peat as a fuel 28
2.2. The peat–air–water system 29
2.3. Water content and heating value 30
3. Water movement in a peat column 31
3.1. Equations of motion and of continuity 31
3.2. Potential energy and the phase-partitioning of water 32
3.3. Liquid and vapour conductivities 33
4. Boundary conditions at the pedon scale 34
4.1. The drainage boundary condition at the bottom 34
4.2. Surface boundary condition for infiltration of rain 35
4.3. Surface boundary condition for isothermal evaporation 36
5. The drying of a layer of milled peat 37
5.1. Modelling milling and harrowing 37
5.2. Numerical treatment 38
5.3. The two phases of evaporation 39
5.4. The role of harrowing 41
6. Conclusions 43
Acknowledgements 44
References 45

Chapter 4. Role of active–passive scalar relationships in evaporation from vegetated surfaces
by Gordon A. McBean 47
Abstract 47
1. Introduction 47
2. Efficiency of evaporation 48
3. Concept of active and passive scalars 50
4. Impact of K_E/K_H ratio on traditional methods of estimating evaporation from vegetated surfaces 53
5. Concluding remarks 55
References 56

Chapter 5. On the weights of precipitation stations
by M. Sugawara 59
Abstract 59
1. The Thiessen polygon method is illogical 59
2. Weights should be determined considering the reliability of observed data 61
3. Weights of precipitation stations should be determined to obtain good results in discharge calculation or flood forecasting 62
4. How to determine the weights of precipitation stations 63
5. The problem 65
6. The singular matrix (A_{ij}) 66
7. The method of principal axis transformation 67
8. Factor analysis 69
9. Orthogonal expansion of the runoff 70

10. The precipitation weights 72
11. The determination of the cut-off point k' 74
References 74

Chapter 6. A Neyman-Scott shot noise model for the generation of daily streamflow time series
by P. S.P. Cowpertwait and P.E. O'Connell 75
Abstract 75
1. Introduction 75
2. Definition of a second order Neyman-Scott shot noise (NSSN) model 76
3. Some moments of the Neyman-Scott shot noise model 78
4. Some special cases 81
5. Fitting the NSSN model to historical daily flows 81
6. Model validation 86
7. Discussion and epilogue 91
Acknowledgements 93
References 94

PART B. LARGE-SCALE HYDROLOGY

Chapter 7. Criteria for a hydrologically sound structuring of large scale land surface process models
by Alfred Becker 97
Abstract 97
1. Introduction 97
2. Scales in hydrology and related categories of hydrological models 98
3. Critical concepts in macroscale hydrological modelling at land surfaces 100
4. Hydrologically sound structuring of macroscale models 102
5. Basic types of areal heterogeneity at land surfaces and assessment of landscape patchiness 105
6. Hierarchy in the areal discretization of land surfaces for modelling 108
7. Criteria to delineate zones of uniform climate forcing 109
References 110

Chapter 8. The construction of continental scale models of the terrestrial hydrological cycle: an analysis of the state of-the-art and future prospects
by L.S. Kuchment 113
Abstract 113
1. Introduction 113
2. Time–space scales and main research directions 114
3. Continental scale hydrological models with present-day space resolution of GCMs (100–300 km) 117
4. Modelling of the terrestrial hydrological cycle for a horizontal scale of 30–100 km . 122
4.1. Basin schematisation 123

4.2. Input to finite elements 123
4.3. Soil moisture transfer and point evapotranspiration 124
4.4. Rainfall excess 124
4.5. Average surface depression interception for each finite element 124
4.6. Surface overland flow 125
5. Opportunities for developing parameterisations of the hydrological cycle for horizontal resolutions of less than 30 km 125
Acknowledgements 126
References 127

Chapter 9. A rainfall-runoff scheme for use in the Hamburg climate model
by L. Dümenil and E. Todini 129
Abstract 129
1. Introduction 129
2. Runoff in the context of general circulation models 130
2.1. Motivation 130
2.2. The ECHAM Model 132
3. Description of the Arno scheme 133
3.1. Derivation of the equations 133
3.2. Modifications for use in the OAGCM 137
4. The simulated hydrological cycle 139
4.1. Precipitation 140
4.2. Local runoff 144
4.3. Soil water 146
4.4. River gauge data 147
5. Conclusions 155
Acknowledgement 155
References 156

Chapter 10. Transient response of a coupled ocean–atmosphere–land surface model to increasing atmospheric carbon dioxide
by S. Manabe, R.J. Stouffer, M.J. Spelman and K. Bryan 159
Abstract 159
1. Introduction 160
2. Model structure 160
3. Numerical experiments 161
4. Temperature change 162
4.1. Annual mean response 162
4.2. Seasonal dependence 163
5. Hydrologic change 166
5.1 Annual mean response 166
5.2. Seasonal dependence 168
6. Summary and concluding remarks 169
References 171

PART C. THE PAST AND THE FUTURE

Chapter 11. Quantitative hydrology in Scandinavia in the 18th century
by Lars Gottschalk 177
Abstract 177
1. Introduction 177
2. Hydrometeorological observations 178
 2.1. Observations of temperature and precipitation 178
 2.2. Evaporation experiments 179
3. Water balance studies 180
4. The origin of springs 182
 4.1. Rain or "underground channels"? 183
 4.2. "Disappearance of water" 185
5. Conclusions 186
References 187

Chapter 12. Hydrology and hydrologists — reflections
by J.E. Nash 191
Abstract 191
1. Introduction 191
2. The alleged failure of engineering hydrology 192
3. Science or technology 193
4. Empirical hydrology 194
5. The two pillars of hydrology 194
6. The real failure of engineering hydrology 195
7. The future for hydrology 196
8. The education and training of hydrologists 196
9. Environmental awareness 198
References 199

Chapter 13. Hydrology and the real world
by J.R. Philip 201
Abstract 201
1. Introduction 201
2. Hydrology and non-linearity 201
3. Natural science and trans-science 202
4. Scientific hydrology and professional practice 203
5. Hydrology and the real world 205
6. Hydrology and faith 205
Acknowledgments 206
References 206

Chapter 14. Water and soil: circulation and risks of pollution
by George Vachaud and Michel Vauclin 209
Abstract 209

1. Introduction 209
2. Presentation of the problem 210
3. Circulation of water in the soil 211
3.1. The microscopic scale of the pore 211
3.2. The macroscopic scale: the laboratory column 211
3.3. Megascopic scale: the landscape 213
4. Transport of dissolved substances 214
5. Conclusion 217
References 217

Chapter 15. Global change, a catalyst for the development of hydrologic science
by Peter S. Eagleson 219
Abstract 219
1. Introduction 220
2. Historical development of hydrologic science 221
3. The geophysical basis of hydrologic science 223
4. The biogeochemical basis of hydrologic science 225
5. Some global change issues in hydrologic science 226
5.1. How do we aggregate the dynamics at various space–time scales in the presence of great heterogeneity? 227
5.2. What are the feedback sensitivities of atmospheric dynamics and climate to changes in landsurface hydrology? 228
5.3. What are the sensitivities of the methane productivity of wetlands to climate change? 229
5.4. What is the physical basis for the observed geographical distribution of the major vegetation types on earth's continents? 230
5.5. Can the dynamics of multistable nonlinear systems suggest new physical insights into the patterns of annual rainfall time series? 231
6. Hydrologic science is a distinct geoscience 233
7. Needed actions 234
Acknowledgements 235
References 235

Chapter 16. The theory of the hydrologic model, or: the struggle for the soul of hydrology
by M.B. Abbott 237
Abstract 237
1. Introduction: maintaining continuity in thinking through a discontinuity in thought 237
2. Navigating through a world of models: the function of the sign 238
3. Understanding through the ordering of signs 241
4. Hydrological rhetoric and its ordering in grammar 244
5. The struggle for the soul of hydrology 247
6. Does the future hold out any hope at all? 251
References 252

Part A. Hydrological processes

Chapter 1

Linear theory of open channel flow

JAROSŁAW J. NAPIÓRKOWSKI

Abstract

The linearized solution of the St. Venant equations is discussed for the general case of any shape of channel and any friction law. The upstream and downstream transfer functions are derived analytically for a channel reach of finite length. For the limiting case of a semi-infinite channel the properties of the upstream transfer function are studied using the cumulants, and amplitude and phase spectra. A number of simplified forms of the St.Venant equations are presented.

1. Introduction

In the unsteady motion of water in canals and rivers, the strategy is followed of proposing simplified models and comparing their performance with field data and with the result of simulations by means of the St. Venant equations. The main aim is to obtain convenient predictive models. Although non-linear systems modelling is already extensively developed, it does not seem useful for stochastic hydrology or for optimal control problems when flow routing in channels must be repeated many times for different scenarios. When the starting point in such a strategy is the linearization of a set of non-linear equations, and moreover the aim is to gain insight into the nature of the complete non-linear problem, the problem then belongs to a very special class. That class has been termed by Polish hydrologists Doogeology in recognition of Prof. Dooge's contribution to the linear theory of open channel flow.

The most important problems in the linear theory of open channel flow are:

(1) The two-point boundary problem in which both upstream and downstream boundary conditions are taken into account.

(2) The downstream problem, i.e. the prediction of the flood characteristics at a downstream section on the basis of a knowledge of the flow characteristics at an upstream section and the hydraulic characteristics of the channel between the two sections.

(3) The steady-state rating curve at a downstream control.

(4) The tributary problem which involves predicting the effect of tributary inflow on conditions in the main channel both upstream and downstream of the point of entry.

(5) The lateral inflow problem in which there is a distributed inflow to the channel reach. The above classification and description applies only to a tranquil (subcritical) flow in which the Froude number is less than one. For rapid (supercritical) flow there is no upstream effect. The present paper concentrates on the first two problems of Doogeology only.

2. Linearization of the St. Venant equations

When only one space dimension is taken into account, the equation of continuity for the unsteady flow in an open channel in the absence of lateral inflow is given by:

$$\frac{\partial Q}{\partial x} + \frac{\partial A}{\partial t} = 0 \tag{1}$$

where $Q(x,t)$ is the discharge, $A(x,t)$ is the cross-sectional area, x is the distance from the upstream boundary, and t is the elapsed time.

If the assumption is made that only acceleration in the direction of motion needs to be taken into account then the equation for the conservation of linear momentum in this direction can be written in terms of the same variables (Dooge et al., 1982):

$$(1 - F^2)g\bar{y}\frac{\partial A}{\partial x} + \frac{2Q}{A}\frac{\partial Q}{\partial x} + \frac{\partial Q}{\partial t} = gA(S_o - S_f) \tag{2}$$

where F is the Froude number, $\bar{y}$ is the hydraulic mean depth (the ratio of the area of flow to the width of the channel at the water surface), S_o is the bottom slope and S_f is the friction slope defined by the equilibrium condition for steady uniform flow.

The friction slope depends on the type of friction law assumed, the shape and roughness of the cross-section, the flow of the section and the area of flow. It can be written in very general form as:

$$S_f = f(A, Q, \text{shape, roughness}) \tag{3}$$

The problem of unsteady open channel flow involves the solution of the above set of non-linear equations subject to given initial conditions and two appropriate boundary conditions. No analytical solution is available and equations (1) and (2) must be solved by some method of numerical approximation. To determine analytically the solution, the St. Venant equations are simplified by considering the first order (linear) variation from a steady-state trajectory.

To compute the linearized equations we make use of expansion of non-linear terms in equation (2) in a Taylor series around the uniform steady state (Q_o, A_o) and limitation of this expansion to the first order increments $Q'(x,t)$, $A'(x,t)$. The resulting equations are (Dooge and Napiórkowski, 1984):

$$\frac{\partial Q'}{\partial x} + \frac{\partial A'}{\partial t} = 0 \tag{4}$$

$$(1-F^2)g\bar{y}_o\frac{\partial A'}{\partial x}+2v_o\frac{\partial Q'}{\partial x}+\frac{\partial Q'}{\partial t}=gA_o\left[-\frac{\partial S_f}{\partial Q}Q'-\frac{\partial S_f}{\partial A}A'\right] \tag{5}$$

where $v_o = Q_o/A_o$ is the average velocity of flow, and the derivatives of the friction slope $S_f(Q,A)$ with respect to discharge Q and the area A on the right-hand side of the equation are evaluated at the reference conditions.

The variation of the friction slope with discharge at the reference condition for all frictional formulae for rough turbulent flow could be taken as:

$$\frac{\partial S_f}{\partial Q}=2\frac{S_o}{Q_o} \tag{6}$$

To express the variation of the friction slope with flow area we define for convenience a parameter m as the ratio of the kinematic wave speed to the average velocity of flow:

$$m=\frac{c_k}{v_o} \tag{7}$$

where c_k is the kinematic wave speed (Dooge and Napiórkowski, 1984):

$$c_k=-\left[\frac{\partial S_f}{\partial A}\right]\bigg/\left[\frac{\partial S_f}{\partial Q}\right]=\frac{dQ}{dA} \tag{8}$$

When equations (6) and (7) are substituted in equation (8) we obtain:

$$\frac{\partial S_f}{\partial A}=2mv_o\frac{S_o}{Q_o} \tag{9}$$

The parameter m is a function of the shape of channel and of area of flow. For a wide rectangular channel with Chezy friction, m is always equal to 3/2 and with Manning friction is always equal to 5/3. For shapes of channel other than rectangles, m will take on different values.

Since equations (4) and (5) are linear first-order equations in two variables, they are equivalent to a single second-order equation in one variable. The most general form of this second-order equation is that obtained by using the unsteady flow potential (Dooge, 1980). This potential can be defined as the function $U'(x,t)$ whose partial derivatives with respect to distance gives minus the perturbation in the area of flow, that is:

$$\frac{\partial U'}{\partial x}=-A'(x,t) \tag{10a}$$

and whose partial derivative with respect to time gives the perturbation from the reference discharge:

$$\frac{\partial U'}{\partial t}=Q'(x,t) \tag{10b}$$

Consequently, the perturbation potential $U'(x,t)$ automatically satisfies the continuity equation (4). When equations (10) are substituted in equation (5) we obtain the dynamic equation for the unsteady flow potential $U'(x,t)$ in the form:

$$(1 - F_{\mathrm{o}}^2)g\bar{y}_{\mathrm{o}}\frac{\partial^2 U'}{\partial x^2} + 2v_{\mathrm{o}}\frac{\partial^2 U'}{\partial x \partial t} + \frac{\partial^2 U}{\partial t^2} = gA_{\mathrm{o}}\left[-\frac{\partial S_{\mathrm{f}}}{\partial A}\frac{\partial U'}{\partial x} + \frac{\partial S}{\partial Q}\frac{\partial U'}{\partial t}\right] \tag{11}$$

Any linear function of the perturbation potential $U'(x,t)$ will also represent a solution of equation (11). Since differentiation is a linear operation, both $A'(x,t)$ and $Q'(x,t)$ will also be governed by an equation of the same form as equation (11). Similarly, any linear combination of solutions is also a solution of the basic linear equation, e.g. the perturbations of the velocity $v'(x,t)$, the surface width of the channel $T'(x,t)$ and the Froude number $F'(x,t)$. The choice of dependent variable in any given problem will be governed largely by the form in which the boundary conditions are given.

3. Solution for finite channel reach

The basic equation (11) to be solved is hyperbolic in form. Accordingly there are two real characteristics defined by:

$$\frac{\mathrm{d}x}{\mathrm{d}t} = c_{1,2} = v_{\mathrm{o}} \pm \sqrt{(g\bar{y})} \tag{12}$$

along which the discontinuities in the derivatives of the solution will propagate. We will consider the basic case of tranquil flow (Froude number less than one) in which $A'(x,t)$ will be prescribed both at the upstream boundary $x = 0$ and at the downstream boundary $x = L$. The use of other boundary conditions does not introduce any new principle.

The problem is to solve equation (11) subject to the double initial condition:

$$A'(x,0) = 0 \qquad \frac{\partial A'}{\partial t} = 0 \tag{13}$$

and subject to the boundary conditions:

$$A'(0,t) = A_{\mathrm{u}}(t) \qquad A'(L,t) = A_{\mathrm{d}}(t) \tag{14}$$

The solution can be sought in terms of the Laplace transform. Equation (11) when transformed to the Laplace transform domain becomes:

$$(1 - F_{\mathrm{o}}^2)g\bar{y}\frac{\mathrm{d}^2\bar{A}}{\mathrm{d}x^2} + \left[-2v_{\mathrm{o}}s + gA_{\mathrm{o}}\frac{\partial S_{\mathrm{f}}}{\partial A}\right]\frac{\mathrm{d}\bar{A}}{\mathrm{d}x} - \left[s^2 + gA_{\mathrm{o}}\frac{\partial S}{\partial Q}s\right]\bar{A} = 0 \tag{15}$$

where $\bar{A}(x,s)$ is the Laplace transform of $A'(x,t)$. Equation (15) is a second-order homogeneous ordinary equation, so the solution can be written in the general form:

$$\overline{A}(x,s) = C_1(s)\exp[\lambda_1(s)x] + C_2(s)\exp[\lambda_2(s)x] \tag{16}$$

where λ_1 and λ_2 are the roots of the characteristic equation for equation (16) and are given by:

$$\lambda_{1,2} = es + f \pm \sqrt{(as^2 + bs + c)} \tag{17}$$

Here the parameters a, b, c, e and f are given in terms of hydraulic variables by the following relationships:

$$a = \frac{1}{g\bar{y}_o(1 - F_o^2)^2} \tag{18a}$$

$$b = \frac{2S_o}{v_o\bar{y}_o}\frac{1 + (m - 1)F_o^2}{(1 - F_o^2)^2} \tag{18b}$$

$$c = \left(\frac{mS_o}{\bar{y}_o}\right)^2 \frac{1}{(1 - F_o^2)^2} \tag{18c}$$

$$e = \frac{v_o}{g\bar{y}_o}\frac{1}{1 - F_o^2} \tag{18e}$$

$$f = \sqrt{(c)} \tag{18f}$$

Having determined the parameters $C_1(s)$ and $C_2(s)$ from the boundary conditions, one can write $\overline{A}(x,s)$ in terms of the $h_u(x,s)$ and $h_d(x,s)$ which may be defined as the Laplace transforms of the responses of the channel reach to delta-function inputs at the upstream and downstream ends respectively. Accordingly we can write:

$$\overline{A}(x,s) = h_u(x,s)\overline{A}_u(s) + h_d(x,s)\overline{A}_d(s) \tag{19}$$

The linear channel responses to an upstream input $h_u(x,s)$ and to a downstream input are given by (Dooge and Napiórkowski, 1987a):

$$h_u(x,s) = \exp[(es + f)x]\frac{\sinh[\sqrt{(as^2 + bs + c)}(L - x)]}{\sinh[\sqrt{(as^2 + bs + c)}L]} \tag{20}$$

$$h_d(x,s) = \exp[-(es + f)(L - x)]\frac{\sinh[\sqrt{(as^2 + bs + c)}x]}{\sinh[\sqrt{(as^2 + bs + c)}L]} \tag{21}$$

The original function in the time domain is determined from the corresponding boundary condition through the relationship:

$$A'(x,t) = h_u(x,t) * A_u(t) + h_d(x,t) * A_d(t) \tag{22}$$

The explicit formulation for the transfer functions $h_u(x,t)$ and $h_d(x,t)$ in the time domain have been obtained by Dooge and Napiórkowski (1987a). The transfer function due to an upstream input is found to have two distinct parts so that we can write:

$$h_u(x,t) = h_u^1(x,t) + h_u^2(x,t) \tag{23}$$

The first part of the solution, which may be termed the head of the wave, is given by:

$$h_u^1(x,t) = \sum_{n=0}^{\infty} \exp(-2nL\alpha_1 - \alpha_2 x)\,\delta(t - nt_o - x/c_1)$$

$$- \sum_{n=1}^{\infty} \exp(-2nL\alpha_1 + \alpha_3 x)\,\delta(t - nt_o - x/c_2) \tag{24}$$

The celerities c_1 and c_2 are defined in equation (12) and the parameters α_1, α_2, α_3 and t_o (functions of the channel parameters) are given by:

$$\alpha_1 = b/2\sqrt{(a)} \tag{25a}$$

$$\alpha_2 = \alpha_1 - f \tag{25b}$$

$$\alpha_3 = \alpha_1 + f \tag{25c}$$

$$t_o = L/c_1 - L/c_2 \tag{25d}$$

It can be seen that the head of the wave moves downstream at the dynamic speed c_1 in the form of a delta function of exponentially declining volume proportional to $\exp(-\alpha_2 x)$. At $x = L$ the delta function is reflected with an inversion of sign and is propagated upstream at the speed c_2 and with a heavier damping factor $\exp[-\alpha_3(L - x)]$. Then it is reflected again at $x = 0$ to move in a downstream direction etc.

The second part of the upstream response, which may be termed the body of the wave, is:

$$h_u^2(x,t) = \sum_{n=0}^{\infty} \exp(-\beta_1 t + \beta_2 x)(h/c_1 - h/c_2)(2nL + x) \cdot$$

$$\frac{I_1\{2h\sqrt{[(t - nt_o - x/c_1)(t + nt_o - x/c_2)]}\}}{\sqrt{[(t - nt_o - x/c_1)(t + nt_o - x/c_2)]}}\,U[t - nt_o - x/c_1]$$

$$- \sum_{n=1}^{\infty} \exp(-\beta_1 t + \beta_2 x)(h/c_1 - h/c_2)(2nL - x) \cdot$$

$$\frac{I_1\{2h\sqrt{[(t - nt_o - x/c_2)(t + nt_o - x/c_1)]}\}}{\sqrt{[(t - nt_o - x/c_2)(t + nt_o - x/c_1)]}}\,U[t - nt_o - x/c_2] \tag{26}$$

where $I_1\{\,\}$ is a modified Bessel function of the first kind, $U[\,]$ is a unit step function, and the remaining parameters are given by:

$$\beta_1 = b/2a \tag{27a}$$

$$\beta_2 = f - be/2a \tag{27b}$$

$$h = \sqrt{(b^2/4 - ac)}/2a \tag{27c}$$

As in the case of the head of the wave, the body of the wave is subject to successive reflection at both the downstream and upstream boundaries but moves and dissipates more slowly than the head of the wave.

For the downstream transfer function, the head of the wave is given by:

$$h_d^1(x,t) = \sum_{n=0}^{\infty} \exp[-2nL\alpha_1 - \alpha_3(L-x)]\,\delta[t - nt_o + (L-x)/c_2]$$
$$- \sum_{n=0}^{\infty} \exp[-(2n+1)L\alpha_1 - fL - \alpha_2 x]\,\delta(t - nt_o + L/c_2 - x/c_1) \tag{28}$$

and is subject to reflection at the two ends of the reach as in the case of $h_u^1(x,t)$. The body of the wave is given by:

$$h_d^2(x,t) = \sum_{n=0}^{\infty} \exp[-\beta_1 t - \beta_2(L-x)](h/c_1 - h/c_2)\,[2nL + (L-x)] \cdot$$
$$\frac{I_1\{2h\sqrt{[(t - nt_o + L/c_2 - x/c_2)(t + nt_o + L/c_1 - x/c_1)]}\}}{\sqrt{[(t - nt_o + L/c_2 - x/c_2)(t + nt_o + L/c_1 - x/c_1)]}}\,U[t - nt_o - x/c_1]$$
$$- \sum_{n=0}^{\infty} \exp[(-\beta_1 t - \beta_2(L-x)](h/c_1 - h/c_2)\,[2(n+1)L + x] \cdot$$
$$\frac{I_1\{2h\sqrt{[(t + nt_o + L/c_1 - x/c_2)(t - nt_o + L/c_2 - x/c_1)]}\}}{\sqrt{[(t + nt_o + L/c_1 - x/c_2)(t - nt_o + L/c_2 - x/c_1)]}}\,U[t - nt_o - x/c_1] \tag{29}$$

Note that:

(1) For the case where there is both an upstream and a downstream boundary condition the unsteady wave motion produced by each of the boundary conditions will be successively reflected at each end of the channel reach and thus will require representation by an infinite series.

(2) Allowance for a downstream boundary condition thus has a double effect, since it produces the reflection of the movement due to the upstream input as well as a direct effect on the channel reach of the downstream boundary condition.

(3) The reflection of the two sets of wave motion at opposite ends of the channel from the point of generation, results in representing each of the linear channel responses by two infinite series, one representing each direction of propagation.

Considering that the modified Bessel function is itself represented by an infinite series, the solution is in the form of a double infinite series which seems too complicated for practical application in river flow forecasting. However, due to heavy damping, only the first few terms of the transfer functions would be required, and the polynomial approximation of the first order modified Bessel function is sufficiently accurate and can be calculated easily (Napiórkowski and Dooge, 1988). It was found moreover, that:

(1) The total volume of the contribution to the input response due to a delta function input at the upstream boundary and a delta function input at the downstream boundary sum to unity for any point in the channel reach.

(2) The rate of convergence in the two infinite series characterizing the response to an upstream input and the two infinite series characterizing the response to a downstream input is the same in all four cases and is equal to $\exp(-2fL)$.

4. The generalized linear channel response

In the present paragraph we will concentrate on the downstream wave which is of primary importance in flood routing. To filter out the downstream wave we can set $L \to \infty$ in equation (20) and deal with the limiting case of downstream flow for a semi-infinite reach (i.e. the case where the downstream control is so distant from the section of interest that the downstream boundary has no influence). In such a case the upstream transfer function in Laplace transform domain and in the time domain is given respectively by:

$$h_u(x,s) = \exp(exs + fx - x\surd(as^2 + bs + c) \tag{30}$$

$$h_u(x,t) = \exp(-\alpha_2 x)\,\delta(t - x/c_1) +$$

$$\exp(-\beta_1 t + \beta_2 x)(h/c_1 - h/c_2)\frac{I_1\{2h\surd[(t - x/c_1)(t - x/c_2)]\}}{\surd[(t - x/c_1)(t - x/c_2)]}\,U[t - x/c_1] \tag{31}$$

which corresponds to use of only the first term in equations (24) and (26). The above expressions are generalized forms (Dooge, 1980; Dooge et al., 1987a) of those obtained by Dooge and Harley (1967a, b) for the special case of a wide rectangular channel with Chezy friction.

4.1. Cumulants of generalized channel response

The use of cumulants (or moments) has been widely used both in unit hydrograph analysis and in flood routing to study the properties of linear responses and to compare the various models proposed for use in representing the linear channel response or the unit hydrograph (Dooge, 1973). If the corresponding cumulants (or moments) of the impulse responses are equal to each other, up to and including N, then both systems respond identically to input signals which are at most Nth order polynomial functions of time.

In establishing a theoretical relationship, it is more convenient to use the cumulants, rather then the moments which are related to them. While the moments are generated by the Laplace transform, the cumulants are generated by the logarithm of the Laplace transform. Accordingly they are given by:

$$k_R[h(x,t)] = (-1)^R \frac{\mathrm{d}^R}{\mathrm{d}s^R} \{\ln[h(x,s)\}_{s=0} \tag{32}$$

The first cumulant is identical with the first moment about the origin; the second cumulant is identical to the second moment about the centre; the third cumulant is identical to the third moment about the centre; but in the case of the higher moments and cumulants this identity does not exist.

It is clear from the above that in case of the linear channel response it is possible to derive the values of the moments or of the cumulants from the solution in the Laplace transform domain given by equation (30) even if the explicit solution in the time domain given by equation (31) is very complicated. Substituting from equation (30) in equation (32) we obtain for the cumulants:

$$k_R[h(x,t)] = (-1)^R \frac{\mathrm{d}^R}{\mathrm{d}s^R} \left\{[-\sqrt{(as^2 + bs + c)} + es + f]x\right\}_{s=0} \tag{33}$$

where the parameters a, b, c, e and f have the values given by equation (18).

The first two cumulants are relatively easy to obtain from implicit equation (33):

$$k_1 = \frac{x}{mv_o} \tag{34a}$$

$$k_2 = \frac{1}{m}[1 - (m-1)^2F_o^2]\left[\frac{x}{mv_o}\right]\left[\frac{\bar{y}_o}{S_o x}\right] \tag{34b}$$

To calculate the cumulants of higher order ($R > 2$) one can use a general expression recently derived by Romanowicz et al. (1988):

$$k_R[h(x,t)] =$$

$$\left[\frac{x}{mv_o}\right]^R \left[\frac{\bar{y}_o}{S_o x}\right]^{R-1} m^{-R} \frac{[1 + (m-1)F_o^2]^R}{1 - F_o^2} \sum_{i=0}^{[R/2]} \gamma(R,i) \left\{\frac{m^2F_o^2}{[1 + (m-1)F_o^2]^2}\right\}^i \tag{35}$$

where the coefficients $\gamma(R,i)$ can be expressed as:

$$\gamma(R,i) = (-1)^i \frac{r!(2R - 2i - 3)!}{(R - i - 2)!(R - 2i)!\,i!}(1/2)^{R-2} \tag{36}$$

The general expression for Rth cumulant as given by equation (35) was found to depend on the time taken for a kinematic wave to traverse the length of the channel $[x/(mv_o)]$, the dimensionless channel length $(S_o x/\bar{y}_o)$, the ratio of the kinematic wave speed to the average velocity of flow (m), and the Froude number of flow at the reference conditions (F_o).

4.2. Amplitude and phase characteristics

In many other branches of geophysics, the properties of systems are discussed in relation to their response to a harmonic rather than impulsive input. This can be done either by means of wave number analysis (Kundzewicz and Dooge, 1989) or by means of frequency analysis. In this paper the second approach is presented.

The amplitude and phase spectra are used as the basis of comparison of various models and of the comparison of models with data. These spectra are also used to gain insight into the characteristics of the model or of the system being simulated.

The amplitude and phase spectra can be derived from the Fourier transform of the system response. This transform is readily obtained by taking only the imaginary part of the argument of the system function:

$$h(x \cdot i\omega) = h(x,s)|_{s=i\omega} = A(x,\omega)\exp[i\varphi(x,\omega)] \tag{37}$$

The quantities $A(x,\omega)$ and $\varphi(x,\omega)$ are called the amplitude characteristic and the phase characteristic, respectively. The frequency characteristics of the generalized downstream channel response are (Dooge et al., 1987b):

$$A(x,\omega) = \exp\left[xf - x\sqrt{\{\sqrt{[b^2\omega^2 + \sqrt{(-a\omega^2 + c)}]} - a\omega^2 + c\}/\sqrt{2}}\right] \tag{38a}$$

$$\varphi(x,\omega) = xe\omega - x\sqrt{\{\sqrt{[b^2\omega^2 + \sqrt{(-a\omega^2 + c)}]} + a\omega^2 - c\}/\sqrt{2}} \tag{38b}$$

where the parameters a, b, c, e and f are defined by equations (18).

It is instructive to examine the form of the amplitude and phase spectra for the limiting values of the frequency ω. For very low frequencies, i.e. very long waves, the upstream boundary condition in the form of a harmonic oscillation:

$$A_u(t) = A_o\cos(\omega t) \tag{39}$$

results in a harmonic oscillation at the point x:

$$A(x,t) = A_o\cos(\omega t - \omega x/mv_o) \tag{40}$$

Note, that the phase velocity of the above wave $c_k = mv_o$ corresponds to the kinematic wave speed.

At the other extreme of very high frequencies, i.e. very short waves, the upstream boundary condition (41) results in a harmonic oscillation:

$$A(x,t) = A_o\exp(-\alpha_2 x)\cos(\omega t - \omega x/c_1) \tag{41}$$

Clearly, the result corresponds to the head of the wave [see equations (24) and (31)] traveling with the phase velocity $c_1 = v_o + \sqrt{(g\bar{y}_o)}$ and attenuation $\exp(-\alpha_2 x)$.

5. Simplified forms of St. Venant equations

A number of models of simplified forms of the complete St. Venant equation given by equation (11) have been proposed in the hydrological literature. If all three of the second-order terms on the left-hand side of that equation are neglected and the dependent variable is A' we obtain:

$$gA_o \left[-\frac{\partial S_f}{\partial A}\frac{\partial A'}{\partial x} + \frac{\partial S}{\partial Q}\frac{\partial A'}{\partial t} \right] = 0 \tag{42}$$

which is equivalent to:

$$c_k \frac{\partial A'}{\partial x} + \frac{\partial A'}{\partial t} = 0 \tag{43}$$

where c_k is kinematic wave speed defined by equation (8). The solution of this linear equation is:

$$A'(x,t) = f(t - x/c_k) \tag{44}$$

which represents a pure translation. The system function of this solution is:

$$H(s) = \exp(-sx/c_k) \tag{45}$$

The first cumulant $k_1 = x/c_k$ reproduces exactly the first cumulant of the complete solution as is given by equation (34a) and all the higher order cumulants are zero. Equation (43) therefore represents an adequate first order approximation and can be used as the basis of a first order analysis of flood waves.

A new-order approximation can be obtained by using equation (44) to approximate two of the terms on the left hand side of equation (11) in terms of the remaining third term. These approximation are:

$$\frac{\partial^2 A'}{\partial x^2} = \frac{1}{c_k^2} f''(\,) \qquad \frac{\partial^2 A'}{\partial x \partial t} = \frac{1}{c^k} f''(\,) \qquad \frac{\partial^2 A'}{\partial t^2} = f''(\,) \tag{46}$$

If the second and third terms are expressed in terms of the first we have:

$$\frac{\partial^2 A'}{\partial x \partial t} = c_k \frac{\partial^2 A'}{\partial x^2} \qquad \frac{\partial^2 A'}{\partial t^2} = c_k^2 \frac{\partial^2 A'}{\partial x^2} \tag{47}$$

Substitution from equation (47) into equation (11) gives diffusion analogy models:

$$D \frac{\partial^2 A'}{\partial x^2} = a \frac{\partial A'}{\partial x} + \frac{\partial A'}{\partial t} \tag{48}$$

where the advective parameter a and the diffusion parameter D are given by:

$$a = c_k \qquad D = 0.5[1 - (m-1)^2 F_o^2] \frac{v_o \bar{y}_o}{S_o} \tag{49}$$

The diffusion analogy model gives the correct value for the second cumulant for any value of F_0. Further comparison for higher cumulants reveals that the diffusion analogy model is identical to the complete solution for the special case of $F_0 = 0$. A number of forms of the diffusion analogy approximation to the linearized St. Venant equation, the solution of the two-point boundary problem were discussed by Dooge and Harley (1967b), Dooge and Napiórkowski (1984, 1987a, b), Dooge et al. (1983).

If the alternative approach is taken of expressing all the second-order terms as cross-derivatives, then we have the relationship:

$$\frac{D}{a}\frac{\partial^2 A'}{\partial x \partial t} + a\frac{\partial A'}{\partial x} + \frac{\partial A'}{\partial t} = 0 \tag{50}$$

where D and a have the same values as defined by equation (49). The solution of this equation has the system function which is identical in form to the system function for the distributed Muskingum model (Strupczewski et al., 1989).

6. Conclusions

The class of problems that can be termed Doogeology, is defined. The results of research concerning two most important problems in Doogeology, namely, the two-point boundary problem and the downstream problem, are presented.

References

Dooge, J.C.I., 1973. Linear theory of Hydrologic systems. U.S. Dep. Agric., Tech. Bull., No. 1468.

Dooge, J.C.I., 1980. Lectures notes on flood routing in channels. Dep. Civ. Eng., University College Dublin.

Dooge, J.C.I. and Harley, B.M., 1967a. Linear theory of open channel flow. Dept. Civ. Eng., University College Cork (unpublished memorandum).

Dooge, J.C.I. and Harley, B.M., 1967b. Linear routing in uniform open channels. Proc. Int. Hydrol. Symp., Fort Collins, Colo., 1: 57–63.

Dooge, J.C.I. and Napiórkowski, J.J., 1984. Effect of downstream control in diffusion routing. Acta Geophys. Pol., 32: 363–373.

Dooge, J.C.I. and Napiórkowski, J.J., 1987a. The effect of the downstream boundary condition in the linearized St. Venant equations. Q.J. Mech. Appl. Math., 40: 245–256.

Dooge, J.C.I. and Napiórkowski, J.J., 1987b. Applicability of diffusion analogy in flood routing. Acta Geophys. Pol., 35: 65–75.

Dooge, J.C.I., Napiórkowski, J.J. and Strupczewski, J.J., 1982. Hydrodynamic derivation of storage parameters of the Muskingum model. J. Hydrol., 54: 371–387.

Dooge, J.C.I., Kundzewicz, Z.W. and Napiórkowski, J.J., 1983. On backwater effects in linear diffusion routing. Hydrol. Sci. J., 28: 391–402.

Dooge, J.C.I., Napiórkowski, J.J. and Strupczewski, J.J., 1987a. The linear downstream response of a generalized uniform channel. Acta Geophys. Pol., 35: 277–291.

Dooge, J.C.I., Napiórkowski, J.J. and Strupczewski, J.J., 1987b. Properties of the generalized downstream channel response. Acta Geophys. Pol., 35: 405–418.

Kundzewicz, J.C.I. and Dooge, J.C.I., 1989. Attenuation and phase shift in linear flood routing. Hydrol. Sci. J., 34(1): 121–139.
Napiórkowski, J.J. and Dooge, J.C.I., 1988. Analytical solution of channel flow model with downstream control. Hydrol. Sci. J., 33: 269–287.
Romanowicz, R.J., Dooge, J.C.I. and Kundzewicz, Z.W., 1988. Moments and cumulants of linearized St.Venant equation. Adv. Water Resour., 11: 92–100.
Strupczewski, W.G., Napiórkowski, J.J. and Dooge, J.C.I., 1989. The distributed Muskingum model. J. Hydrol., 111: 235–257.

Chapter 2

Analytical and numerical modelling of unsaturated flow

Q.J. WANG

Abstract

Analytical solutions of simplified forms of the Richards' equation are discussed. Some limiting forms of solutions that bound solutions for real soils are examined. It is shown that the solution of the Richards' equation by full linearisation provides a reasonably good approximation for infiltration when rainfall is of relatively high intensity, but results in too fast surface drainage during evaporation. In numerical schemes for solving the complete non-linear Richards' equation, on the other hand, small times steps have to be used to guarantee stability when surface flux is large, but relatively large time steps can be used when surface flux is small.

A hybrid scheme for solving the Richards' equation is proposed for future research. The solution would consist of a superposition of the redistribution solution for the complete non-linear equation under zero surface flux and the linear perturbation solution for the given surface flux, so as to combine the strength of both numerical and analytical approaches.

1. Introduction

General circulation models provide a means for medium and long range weather forecasting and are a principal tool for studying possible climate change which is one of the most important global environmental issues. The treatment of land surface fluxes in general circulation models is one of the key areas where improved simulation is required.

Professor Dooge is one of the earliest hydrologists who identified this important research area and the necessity of dialogue between climate modellers and hydrologists. His wide and thorough analysis of scale problems in hydrology pointed to the complexity of hydrological processes involved in the scale of general circulation models and to a number of possible approaches, including partial analysis, for tackling it. In the past ten years, Professor Dooge and his research team in the Centre for Water Resources Research, University College Dublin, have been actively involved in the investigation and development of physically based hydrological modules for general circulation models. His invaluable contribution to climate modelling research is reflected by his large number of publications. In the past ten years, he has published over a dozen papers as sole author (e.g., Dooge, 1982, 1984, 1986, 1990, 1991) and has been a joint author of more than a dozen published papers and project reports (e.g., Dooge and O'Kane, 1986; Romanowicz et al., 1990a, b; Kühnel et al., 1991), in the area of climate modelling.

As part of this on going research, this paper addresses the problem of modelling soil moisture movement based on the Richards' equation. The properties of analytical and numerical solutions are investigated, and a hybrid solution scheme is proposed for future research.

2. Physics of soil moisture flow

The equation of continuity for vertical flow in a soil column is:

$$\frac{\partial \theta}{\partial t} + \frac{\partial q}{\partial z} = 0 \tag{1}$$

where $\theta(z,t)$ is the local volumetric moisture content, $q(z,t)$ is the face velocity taken vertically downward, z is the depth below the surface, and t is the elapsed time. The equation of motion for vertical movement of soil moisture is:

$$q = -K\left(\frac{\partial \psi}{\partial z} - 1\right) \tag{2}$$

where K is the effective hydraulic conductivity, and $\psi(z,t)$ is the matric potential which is negative in the unsaturated zone. The above continuity and flow equations can be combined to give a single prognostic equation:

$$\frac{\partial \theta}{\partial t} = \frac{\partial}{\partial z}\left[K\left(\frac{\partial \psi}{\partial z} - 1\right)\right] \tag{3}$$

which is often referred to as the Richards' (1931) equation. If the matric potential ψ and the hydraulic conductivity K are both taken as single valued functions of moisture content θ, we can define hydraulic diffusivity D as:

$$D = K\frac{\mathrm{d}\psi}{\mathrm{d}\theta} \tag{4}$$

and combine equations (3) and (4) in the single equation:

$$\frac{\partial \theta}{\partial t} = \frac{\partial}{\partial z}\left[D\frac{\partial \theta}{\partial z} - K\right] \tag{5}$$

for the soil moisture content θ.

Alternatively one can define the specific water capacity C as:

$$C = \frac{\mathrm{d}\theta}{\mathrm{d}\psi} \tag{6}$$

and write equations (3) and (6) as a single prognostic equation for the soil moisture

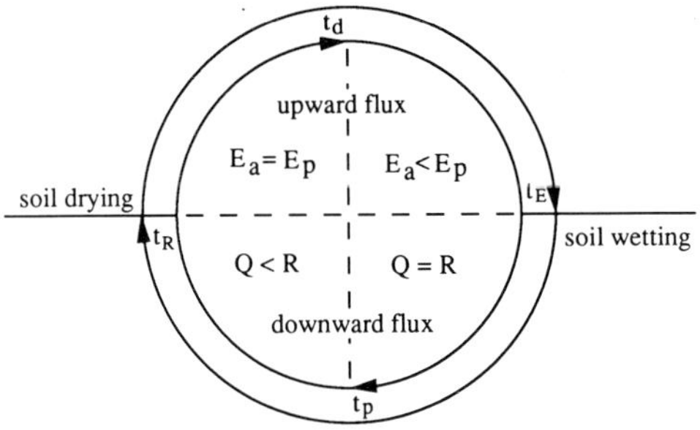

E_p - potential evaporation
E_a - actual evaporation
R - rainfall rate
Q - infiltration rate

t_d - time to desaturation
t_E - total duration of drying period
t_p - time to surface ponding
t_R - total duration of rainfall period

Fig. 1. Four phases of soil–atmosphere interchange of moisture during a wetting and drying cycle.

potential ψ:

$$C\frac{\partial \psi}{\partial t} = \frac{\partial}{\partial z}\left[K\left(\frac{\partial \psi}{\partial z} - 1\right)\right] \tag{7}$$

which has the same form as the Fokker-Planck equation used in certain heat conduction problems (Carslaw and Jaeger, 1946).

For natural soils, functions of the relevant properties — matric potential $\psi(\theta)$, hydraulic conductivity $K(\theta)$ or $K(\psi)$, hydraulic diffusivity $D(\theta)$ and specific water capacity $C(\psi)$ — are all highly non-linear (e.g., Philip, 1969; Mualem, 1976; van Genuchten and Nielsen, 1985). Consequently the basic equations (5) and (7) are also highly non-linear.

Two types of surface boundary conditions are relevant. One is an imposed flux boundary condition associated with atmosphere controlled infiltration or evaporation. The other is a concentration boundary condition associated with soil controlled infiltration or evaporation. Switching from one type of boundary condition to the other is an important aspect in studying the complete cycle of moisture exchange between atmosphere and soil. The four phases of soil–atmosphere interaction are shown in Figure 1.

3. Reduced form of Richards' equation

It is sometimes convenient to rewrite equation (5) in terms of reduced variables leaving the equation dimensionless. This is achieved by setting:

$$\Theta = \frac{\theta - \theta_r}{\theta_s - \theta_r} \qquad \text{and} \qquad K^* = \frac{K - K_r}{\theta_s - \theta_r} \tag{8}$$

where θ_s is the saturated soil moisture content, θ_r is the residual or air–dry soil moisture content, $K_s = K(\theta_s)$ is the hydraulic conductivity at θ_s, and $K_r = K(\theta_r)$ is the hydraulic conductivity at θ_r. The time and depth variables can be converted to dimensionless form:

$$\tau = \frac{K_s^{*2} t}{D^*} \quad \text{and} \quad \zeta = \frac{K_s^* z}{D^*} \tag{9}$$

where K_s^* is the reduced saturated hydraulic conductivity, and D^* is a reference hydraulic diffusivity. The Richards' equation (5) can be written in dimensionless form:

$$\frac{\partial \Theta}{\partial \tau} = \frac{\partial}{\partial \zeta}\left[D/D^* \frac{\partial \Theta}{\partial \zeta}\right] - \frac{\mathrm{d}(K^*/K_s^*)}{\mathrm{d}\Theta}\frac{\partial \Theta}{\partial \zeta} \tag{10}$$

For an imposed surface flux q^0, the surface boundary condition becomes:

$$\frac{Q^0}{K_s^*} = -D/D^* \frac{\partial \Theta}{\partial \zeta} + K^*/K_s^* \qquad \text{at } \zeta = 0 \tag{11}$$

where:

$$Q^0 = \frac{q^0 - K_r}{\theta_s - \theta_r} \tag{12}$$

This imposed flux boundary condition becomes inoperable when surface soil is saturated or air-dry, and a constant concentration boundary condition then applies:

$$\Theta = \Theta^0 \qquad \text{at } \zeta = 0 \tag{13}$$

The latter type of boundary condition also applies to the case of instantaneous ponding.

4. Closed form solutions for constant D and linear K^* model

A number of closed form analytical solutions of Richards' equation are available. Although most of these solutions are based on the assumption of highly simplified function forms of $K(\theta)$ and $D(\theta)$, they are useful in giving insight into the physical problem and providing both approximate solutions and limiting solutions that bound the solutions for real soils. Table 1 gives a list of the available closed form analytical solutions.

Under the assumption that D is equal to an effective constant and K^* is a linear function of Θ:

$$K^* = K_s^* \Theta \tag{14}$$

closed form analytical solutions are available for a semi-infinite homogeneous soil column with uniform initial condition:

$$\Theta = \Theta_0 \qquad \text{at } \tau = 0 \tag{15}$$

TABLE 1

References to closed form analytical solutions of Richards' equation for various functions of diffusivity D and hydraulic conductivity K

	constant K	linear K	nonlinear K
delta function D	Green and Ampt (1911) Mein and Larson (1973)		
constant D	Carslaw and Jaeger (1946) Braester (1973)	Philip (1966) Braester (1973)	Philip (1974) Clothier et al. (1981)
Fujita D	Fujita (1952) Knight and Philip (1974)	not solved Rogers et al. (1983)	not solved Sander et al. (1988)

Author's name on top: concentration boundary condition; author's name at bottom: flux boundary condition.

For an instantaneous ponding boundary condition, the surface moisture flux has been found as (Philip, 1966, 1969):

$$\frac{Q^0}{K_s^*} = \Theta_0 + (1 - \Theta_0)\left[\sqrt{\frac{1}{\pi\tau}}\exp\left(-\frac{\tau}{4}\right) + \frac{1}{2}\operatorname{erfc}\left(-\sqrt{\frac{\tau}{4}}\right)\right] \tag{16}$$

For a constant flux boundary condition, the surface moisture content has been found as (Braester, 1973):

$$\Theta(0,\tau) = \frac{Q^0}{K_s^*}\phi(\tau) + \Theta_0[1 - \phi(\tau)] \tag{17}$$

$$\phi(\tau) = 1 + \sqrt{\frac{\tau}{\pi}}\exp\left(-\frac{\tau}{4}\right) - \left(1 + \frac{\tau}{2}\right)\operatorname{erfc}\left(\sqrt{\frac{\tau}{4}}\right) \tag{18}$$

And for a quantised flux boundary condition, the surface moisture content has been found as (Wang and Dooge, 1991):

$$\Theta(0, i\Delta\tau) = \sum_{j=1}^{i}\frac{Q_{i-j+1}^0 - Q_{i-j}^0}{K_s^*}\phi_j + \Theta_0(1 - \phi_i) \tag{19}$$

$$\phi_j = 1 + \sqrt{\frac{j\Delta\tau}{\pi}}\exp\left(-\frac{j\Delta\tau}{4}\right) - \left(1 + \frac{j\Delta\tau}{2}\right)\operatorname{erfc}\left(\sqrt{\frac{j\Delta\tau}{4}}\right) \tag{20}$$

or:

$$\Theta(0, i\Delta\tau) = \sum_{j=1}^{i} \frac{Q^0_{i-j+1}}{K^*_s}\omega_j + \Theta_0(1 - \phi_i) \tag{21}$$

$$\omega_j = \phi_j - \phi_{j-1} \qquad \text{and} \qquad \phi_0 = 0 \tag{22}$$

where $\Delta\tau$ denotes the discrete time interval.

The last solution can be applied to both atmosphere and soil controlled infiltration or evaporation. When infiltration or evaporation is atmosphere controlled, Q^0_i is the rate of rainfall or potential evaporation. If the surface moisture content $\Theta(0, i\Delta\tau)$, calculated by using the potential rate as Q^0_i, is greater than 1 or less than 0, the corresponding infiltration or evaporation becomes soil controlled. The actual Q^0_i can be found by setting in equation (21) $\Theta(0, i\Delta\tau) = 1$ for soil controlled infiltration or $\Theta(0, i\Delta\tau) = 0$ for soil controlled evaporation. Switching between atmosphere control and soil control can thus be easily made, and surface moisture content be rapidly calculated.

An example is shown in Figure 2 to demonstrate the use of the last solution. In the initial stage of a rainstorm, infiltration is at the potential rate, and the surface soil moisture content increases rapidly. When the surface soil is saturated, the infiltration becomes soil controlled with a decreasing rate. After the rainstorm, evaporation takes place at the potential rate, and the surface soil moisture content decreases due to both uptake of soil moisture by evaporation and drainage by gravity. When the surface soil is air-dry, the evaporation becomes soil controlled with a decreasing rate.

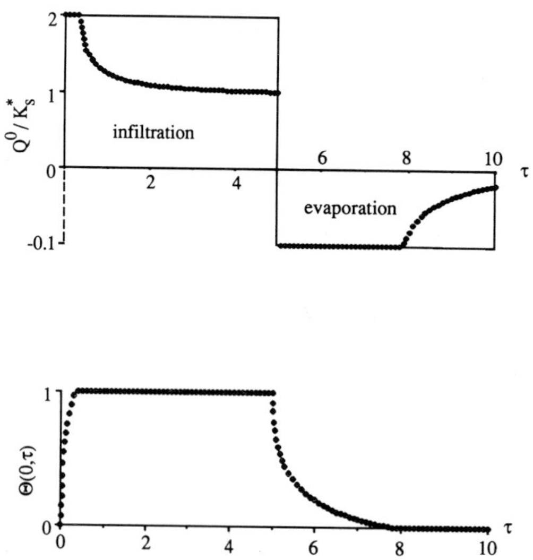

Fig. 2. Example results of constant D and linear K^* model for varied suface flux.

5. Effect of assumptions on D and K^* functions

For natural soils, the diffusivity D falls between constant and delta function forms. A comparison of solutions for these two limiting forms of D may provide some general indication about the accuracy of using an effective constant D (Philip, 1966; Wang and Dooge, 1991).

The following assumptions have been made for such a comparison: (a) both the constant and delta D functions have the same average value; (b) the hydraulic conductivity is constant $K^* = K_s^*$; and (c) the initial soil column is air-dry. The reference hydraulic diffusivity D^* used in equation (9) for deriving reduced variables has been taken equal to the average diffusivity. Figure 3 shows results for infiltration under a typical value of rainfall $P/K_s^* = 10$. It can be seen that the band bounded by the two solutions is relatively narrow, indicating that the assumption of a constant D is adequate here.

The same methodology can be used to examine the effect of assumption on the K^* function. The limiting cases considered this time are: $K^* = 0$ and $K^* = K_s^*$. The following assumptions have been made: (a) the hydraulic diffusivity is equal to an effective constant value; and (b) the initial soil column is air-dry. In deriving the reduced variables of equation (9), the reference hydraulic diffusivity D^* has been taken equal to the effective constant diffusivity, and the same saturated hydraulic conductivity K_s^* has been used for both cases even though it is not a model parameter for the case of constant zero hydraulic conductivity. Figure 4 shows results for infiltration, again under a typical value of rainfall $P/K_s^* = 10$. The two solutions form a band which can be further reduced since $Q^0/K_s^* \geq 1$. The solution for linear K^* should fall into this reduced band. As the band is relatively narrow, the assumption of linear K^* seems to be adequate here.

For evaporation, the depletion of surface soil moisture is due to both drainage by gravity and uptake by air. For natural soils, the linear K^* function, when fitted to values of hydraulic conductivity at air-dry and full saturation, will give unreasonably high value of hydraulic conductivity for moisture content lying in between air-dry and full saturation. This means that the use of the linear K^* function will produce too fast surface drainage. This is strongly evidenced from Figure 5 which shows that

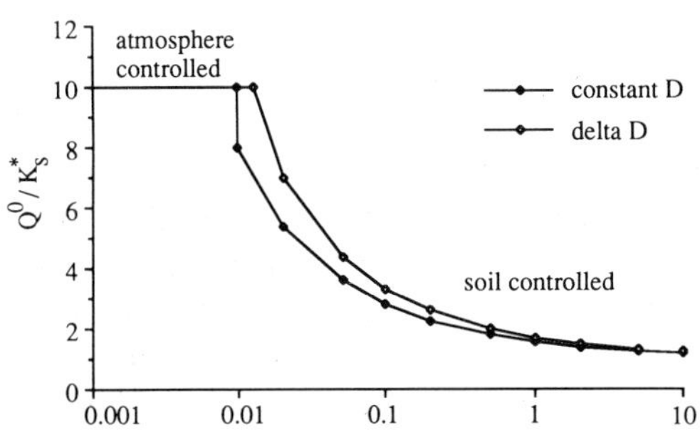

Fig. 3. Effect of D function on surface flux.

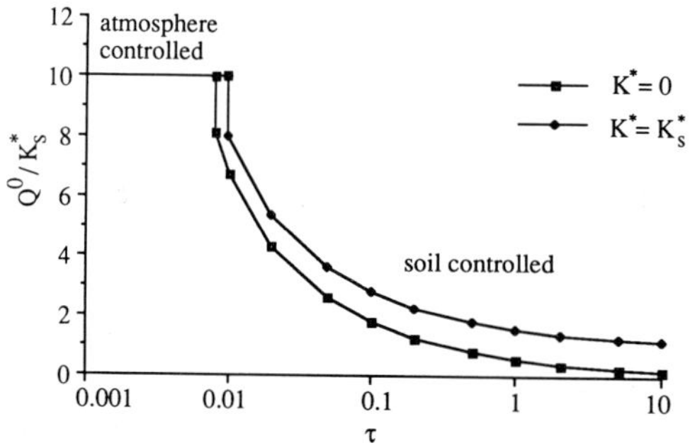

Fig. 4. Effect of K^* function on surface flux.

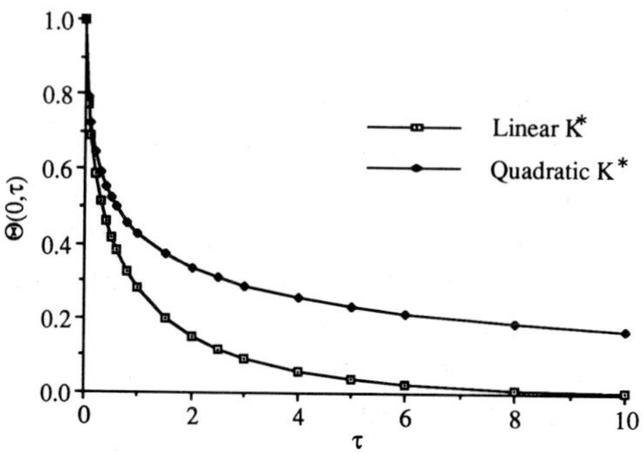

Fig. 5. Effect of K^* function on surface moisture content during redistribution.

the linear K^* function results in much faster surface drainage than the quadratic function $K^* = K_s^*\Theta^2$ during redistribution from an initially saturated soil column under zero surface flux. A constant D function has been assumed in both cases, and the redistribution solution for the constant D and quadratic K^* functions has been obtained by using results of Clothier et al. (1981). Even the quadratic K^* function would predict too high value for hydraulic conductivity when soil becomes relatively dry.

6. A proposed hybrid solution scheme for future research

A number of numerical schemes have been used in the past to solve the Richards' equation (e.g., Haverkamp et al., 1977; Haverkamp and Vauclin, 1981; Campbell, 1985; Thomas and Rees, 1991). As the Richards' equation is highly non-linear, time steps used in these numerical schemes have to be small to guarantee stability when surface flux is large, leading to expensive computation which is impractical for some applications such as for general circulation models. However, relatively large time steps can be used in these numerical schemes when surface flux is small.

On the other hand, the analysis in the last section indicates that the analytical solutions of the constant D and linear K^* model provide reasonably good approximation for infiltration when rainfall is of relatively high intensity but result in too fast surface drainage for evaporation or redistribution. Thus, the numerical and analytical approaches are seen to complement the weakness of one with the strength of the other, and it should prove fruitful to formulate a hybrid algorithm which combines the strength of both approaches.

The following procedures are tentatively proposed. At each time step, the solution for redistribution, i.e., zero surface flux, is found by solving numerically the complete non-linear Richards' equation. The solution of the perturbation from the redistribution profile is then found analytically using an effective constant D and linear K^* model, the perturbation being equal to the actual surface flux. Finally, the overall solution is the superposition of the redistribution solution and perturbation solution.

The proposed procedures would combine the strength of both numerical and analytical approaches, and relatively large time steps could be used throughout, while retaining reasonably good accuracy. Preliminary results from computational experiments seem to support this proposition, but substantial work is still to be carried out.

References

Braester, C., 1973. Moisture variation at the soil surface and the advance of the wetting front during infiltration at constant flux. Water Resour. Res., 9: 678-694.

Campbell, G.S., 1985. Soil Physics with BASIC — Transport Models for Soil Plant Systems. Elsevier, Amsterdam, 150 pp.

Carslaw, H.S. and Jaeger, J.C., 1946. Conduction of Heat in Solids. Oxford University Press, Oxford.

Clothier, B.E., Knight J.H. and White, I., 1981. Burger's equation: application to field constant-flux infiltration. Soil Sci., 132: 255–261.

Dooge, J.C.I, 1982. Parametrization of hydrologic processes. In: P.S. Eagleson (Editor), Land Surface Processes in Atmospheric General Circulation Models. Cambridge University Press, Cambridge, pp. 243–288.

Dooge, J.C.I., 1984. Modelling the behaviour of water. In: E.R.C. Reynolds and F.B. Thompson (Editors), Forests, Climate and Hydrology: Regional Impacts. UNU Workshop, Oxford, March, 1984. United Nations University, Tokyo, 1988, pp. 128–161.

Dooge, J.C.I., 1986. Effects of CO_2 increases on hydrology and water resources. In: R. Fantechi and A. Ghazi (Editors), Carbon Dioxide and other Greenhouse Gases: Climatic and Associated Impacts. Proc. EC Symposium, Brussels, November, 1986. Kluwer, Dordrecht, 1989, pp. 204–213.

Dooge, J.C.I., 1990. Hydrologic models and climate change. Keynote Address to AGU Chapman Conf. on Hydrological Aspects of Global Climate Change, Lake Chelan, Washington, 1990. J. Geophys. Res., Atmos., 1992, 97(D3): 2675–2676,

Dooge, J.C.I., 1991. Sensitivity of runoff to climate change: a Hortonian approach. Robert E. Horton Memorial Lecture, Annu. Meet. Am. Meteorol. Soc., New Orleans, La., January, 1991.

Dooge, J.C.I. and O'Kane, J.P.J., 1986. The role of soil moisture in climate modelling. Final Project Report, EEC/NBST Project CLI-038-EIR(H). Centre for Water Resources Research, University College Dublin.

Fujita, H., 1952. The exact pattern of concentration dependent diffusion in a semi-infinite medium, Part II. Textile Res. J., 22: 823–827.

Green, W.H. and Ampt, G.A., 1911. Studies on soil physics, I. Flow of air and water through soils. J. Agric. Sci., 4: 1–24.

Haverkamp, R. and Vauclin, M., 1981. A comparative study of three forms of the Richard equation used for predicting one-dimensional infiltration in unsaturated soil. Soil Sci. Soc. Am. J., 45: 13–20.

Haverkamp, R., Vauclin, M., Touma, J., Wierenga P.S. and Vachaud, G., 1977. A comparison of numerical simulation models for one-dimensional infiltration. Soil Sci. Soc. Am. J., 41: 285–294.

Knight, J.H. and Philip, J.R., 1974. Exact solutions in nonlinear diffusion. J. Eng. Math., 8: 219–227.

Kühnel, V., Dooge, J.C.I., O'Kane J.P.J. and Romanowicz, R.J., 1991. Partial analysis applied to scale problems in surface moisture fluxes. Surv. Geophys., 12: 221-241.

Mein, R.G. and Larson, C.L., 1973. Modelling infiltration during a steady rain. Water Resour. Res., 9: 384–394.

Mualem, Y., 1976. A catalogue of the hydraulic properties of unsaturated soils. Research Project Report 442, Technion Institute of Technology, Haifa.

Philip, J.R., 1966. A linearization technique for the study of infiltration. In: R.E. Rijtema and H. Wassink (Editors), Water in the Unsaturated Zone. Proc. IAHS/UNESCO Symp., UNESCO, Paris, 1: 471–478.

Philip, J.R., 1969. Theory of infiltration. In: V.T. Chow (Editor), Adv. Hydrosci., 5: 216–296.

Philip, J.R., 1974. Recent progress in the solution of nonlinear diffusion equation. Soil Sci., 117: 257–264.

Richards, L.A., 1931. Capillary conduction of liquids through porous mediums. J . Phys., 1: 318–333.

Rogers, C., Stallybrass M.P. and Clements, D.L., 1983. On two-phase infiltration under gravity and with boundary infiltration: An application of a Bäcklund transformation. Nonlinear Anal. Theory Math. Appl., 7: 785–799.

Romanowicz, R.J., Dooge J.C.I. and O'Kane, J.P.J., 1990a. Partial analysis at particle and pedon scale. EV4C-0076-IRL Project Rep., Centre for Water Resources Research, University College Dublin.

Romanowicz, R.J., Dooge J.C.I. and O'Kane, J.P.J., 1990b. Partial analysis at field and biome scale. EV4C-0076-IRL Project Rep., Centre for Water Resources Research, University College Dublin.

Sander, G.C., Parlange, J.-Y., Kühnel, V., Hogarth, W.L., Lockington D. and O'Kane, J.P.J., 1988. Exact nonlinear solution for constant flux infiltration. J. Hydrol., 97: 341–346.

Thomas, H.R. and Rees, S.W., 1991. A comparison of field-monitored and numerically predicted moisture movement in unsaturated soil. Int. J. Numer. Anal. Methods Geomech., 15: 417–431.

van Genuchten, M.Th. and Nielsen, D.R., 1985. On describing and predicting the hydraulic properties of unsaturated soils. Ann. Geophys., 3(5): 615–628.

Wang, Q.J. and Dooge, J.C.I., 1991. Limiting cases of land surface fluxes. XVIth General Assembly of European Geophysical Society, Wiesbaden, April, 1991. J. Hydrol. (submitted).

Chapter 3

The hydrology of milled peat production

J. PHILIP O'KANE

Abstract

Dried peat is used as a fuel in a number of countries. Drawing on concepts in hydrology and soil physics we present a physically based theory of water movement in unsaturated peat. We apply the theory to the simulation of the drying of a layer of milled peat during a typical production period of four days. The control variables are: (1) the depth to the drained water table, (2) the depth of the milled layer, and (3) the timing and frequency of harrowing.

Various ways of characterising peat water content are discussed. Volume wetness is chosen as the natural state variable and related to the heating value of peat as a fuel. As an initial approximation the peat matrix is assumed to be rigid. Extensions are possible for the non-rigid case.

We derive the differential equation for liquid and water vapour flow in a column of unsaturated peat of any thickness, subject to boundary conditions (a) at the surface for infiltration of rain or evaporation, and (b) at the bottom of the column, for drainage to a water table or field drains.

Two distinct phases of drying are predicted by the theory. In the first phase, evaporation is driven by atmospheric conditions. In the second phase, evaporation is controlled by properties of the peat deposit. The switch is sudden and evaporation declines rapidly, the evaporating surface moving downwards into the peat with the development of a vapour bottleneck.

The function of harrowing is to return evaporation to the first phase, in order to capture as much wind and solar energy as possible.

1. Introduction

Hydrology, as a science, has developed largely in response to questions arising from the technology of water as a resource. In this paper we consider the technology of removing water from peat in order to enhance its value as a fuel. First we describe the problem, then the technology.

Peat is partially decayed organic matter, mainly of plant origin, on the surface of the ground. It is the precursor of coal, oil and gas. A peat bog is "a giant inefficient compost heap, which, left to its own devices, can never complete its rotting process, because it is super-saturated with water. It is a gigantic store of energy and chemicals that can be an investment for all our futures and a repository of information relating to all our pasts" (Bellamy, 1986).

Deposits of peat may be found throughout the temperate zone in North America and Europe, and also in the tropics. In Russia, Finland and Ireland, large quantities of peat are "won" from nature as a fuel; originally it was cut by hand and dried in the open air; machines do the mechanical work now. To win peat as a fuel, the free renewable energies: gravity, sun and wind, must be used to remove water from the organic matter, before it will burn. Consequently, we can release the "gigantic store of energy", only if we capture a matching quantity of solar and gravitational energy. This is an important constraint on the economic production of peat as a fuel.

A number of other concerns may also constrain production in the future. The first is the need to conserve the best examples of bog and fen ecosystems, since some countries, notably the Netherlands, have already lost almost all their peatbogs. The second is the fear that the pronounced change in climate predicted for northern latitudes, may accelerate the composting of peat, releasing still more "greenhouse" gases: carbon dioxide and methane, from the "gigantic store of chemicals".

The review by Dooge (1975) should be consulted on the general hydrology of bogs and fens.

In this paper, we explore mathematically the hydrology of one technology for winning peat: milling and harrowing. The theory of drainage, drying and rewetting, of layers of milled and harrowed peat, under natural conditions, allows us to simulate some important aspects of this technology. The object is to remove water more efficiently from peat, i.e. to capture more alternative energy. We can profit from this knowledge in two ways. For the conservationist, it means we can mill less peat, while producing the same output of fuel energy. On the other hand, for the developer, it means more fuel energy for the same degree of milling. The prospect of carbon and methane taxes will promote the conservation of peat and its more efficient production.

Peat bogs are prepared for milling, (a) by sinking long, closely spaced, parallel, drains, (b) waiting for the drainage to consolidate the peat, and (c) mechanically removing the layer of mossy plants. The bare peat surface is milled to a desired depth with a drum encrusted with pins set to the chosen length. The spread of milled peat may be harrowed mechanically to promote drying by exposing wetter peat to the atmosphere.

2. The peat–water system

2.1. *Peat as a fuel*

Peat in its natural state is largely composed of water. We can consider a notional experiment to find the heating value of undisturbed peat as a fuel using an appropriate calorimeter. The heating value obtained from the experiment would be close to zero or even negative. In this condition the organic material has a dry calorific value roughly equal to the latent heat required to vapourise the water which is present. In section 2.3 we provide a simple model which relates heating value to water content.

For peat to exhibit a large and positive net heating value, free "alternative" energies — wind, sun, and gravity — must be used to remove the water. Energy subsidies embodied in a production technology are justified, when they amplify the capture of free energy.

Production of peat in the open air must wait for and capitalise on periods of sustained evaporation. These periods are interrupted by rain. When another dry period starts, drainage is initially more important than evaporation in removing infiltrated rain–water. At low water content, capillary and adhesive forces in the peat matrix are strong enough to resist gravity. Gradually, evaporation into the atmosphere removes the remaining water and the peat surface may become air-dry.

Clearly, a physical theory of water movement in peat in both the liquid and vapour phase, under ambient atmospheric conditions, is of central importance in the production of peat as a fuel. The theory can be used: (a) to optimise current production technologies, (b) to design and evaluate new technologies, and (c) to design and implement feedback or feedforward controllers.

2.2. The peat–air–water system

Following the notation of soil physics (Campbell, 1985), we define several variables that describe the three-phase peat–air–water system. In a sample of peat, the total mass (kg), M_t, is divided between the mass of air, M_a, which is usually ignored, the mass of water, M_w, and the mass of solids, M_s. The total volume (m^3), V_t, is divided between the volume of air, V_a, the volume of water, V_w, and the volume of solids, V_s. Using these definitions we may define the quantities:

– peat particle density (kg m^{-3})

$$\rho_s = \frac{M_s}{V_s} \tag{1}$$

– dry bulk density (kg m^{-3})

$$\rho_b = \frac{M_s}{V_t} \tag{2}$$

– total fluid porosity (m^{-3} m^{-3})

$$\phi_f = \frac{(V_a + V_w)}{V_t} = \frac{(V_t - V_s)}{V_t} = \left(1 - \frac{\rho_b}{\rho_s}\right) \tag{3}$$

There are a number of ways of characterising the water content of peat:

– mass wetness or mass of water per unit mass of peat solids

$$w = \frac{M_w}{M_s} \tag{4}$$

– total mass wetness or mass of water per unit total mass

$$w_t = \frac{M_w}{M_t} \tag{5}$$

– volume wetness or volumetric water content or equivalent depth of water per unit area

$$\theta = \frac{V_w}{V_t} \tag{6}$$

If the peat matrix is rigid, the dry bulk density ρ_b is constant, and we may define a saturation or maximum volume wetness of the peat equal to the total fluid porosity $\theta_s = \phi_f$.

We include water vapour in the condensed water for simplicity (Rose, 1966). The total mass of water may be partitioned between the liquid and vapour phases when required as shown in section 3.2.

2.3. *Water content and heating value*

From the definitions, mass wetness and total mass wetness are related as follows:

$$w = \frac{w_t}{(1 - w_t)} \tag{7}$$

$$w_t = \frac{w}{(1 + w)} \tag{8}$$

We may express the heating value of peat (H = heat per kg of M_t) at a water content w_t as:

$$H = C_f(1 - w_t) - L_w w_t = C_f - (L_w + C_f)w_t \tag{9}$$

where L_w is the heat of vaporisation of water at standard atmospheric pressure (2256 kJ kg^{-1} of M_w) and C_f is the calorific value of dry peat solids (typically 22,000 kJ kg^{-1} of M_s). At values of $w_t > 0.9$ the heating value of peat is negative. A typical target value in harvesting peat is $w_t = 0.55$, $H \approx 8660$ kJ kg^{-1} of M_t.

Mass wetness w may be expressed in terms of volume wetness q, if we know the dry bulk density $\rho_b(\theta)$ at θ. In this case:

$$w = \frac{\theta \rho_w}{\rho_b(\theta)} \tag{10}$$

where ρ_w is the density of water (M_w/V_w).

The movement of water through the pore space of peat, due to evaporation, drainage and infiltration of rain, is central to our theory. This is most easily expressed in terms of the volume wetness θ. Hence the natural state variable is θ. H is a direct non-linear function of the state variable and may be found by substituting θ in (10), and using (8) and (9).

Variations in dry bulk density with depth or with water content can be accommodated as extensions of the theory. For simplicity only variations with depth are included in this paper.

3. Water movement in a peat column

3.1. Equations of motion and of continuity

The two equations that govern the vertical movement of water in an unsaturated rigid column of bare peat at the pedon scale ($\approx$1 m) are taken from soil-physics (O'Kane et al., 1987). These are:

(1) a water balance, that states that water mass is conserved; and

(2) a dynamic equation, with liquid and vapour water flux proportional to the gradient in peat water potential.

If z is a position coordinate (m) measured vertically downwards from the peat surface and t is time (s), we may write these equations as follows:

$$\rho_w \frac{\partial \theta}{\partial t} = -\frac{\partial (f_l + f_v)}{\partial z} \qquad 0 \leq \theta \leq \theta_s < 1 \tag{11}$$

$$f_l = -k_l \frac{\partial \psi}{\partial z} \tag{12a}$$

$$f_v = -k_v \frac{\partial \psi}{\partial z} \tag{12b}$$

where f_l is the flux density of liquid water (kg m^{-2} s^{-1}) in the direction of positive z, and is proportional to the gradient in the peat water potential $\partial\psi/\partial z$ (Buckingham, 1907); f_v is the flux density of water vapour (kg m^{-2} s^{-1}), and is also proportional to the gradient in the peat water potential $\partial\psi/\partial z$ (Philip, 1955). The liquid and vapour conductivities k_l and k_v are functions of water content.

Different authors define the potential energy of peat water at the pedon scale ($\approx$1 m), in one of three equivalent ways: as energy per unit mass, as energy per unit weight or as energy per unit volume. The gradient of ψ has units of force per unit mass of water, N kg^{-1}, when ψ is measured as J kg^{-1}.

The potential ψ is assumed to be the sum of several component potentials with respect to a common reference state. The most important of these are the gravitational potential ψ_g and the matric potential ψ_m. Hence:

$$\psi = \psi_m + \psi_g = \psi_m - gz \qquad \psi_m \leq 0 \tag{13}$$

where g is the acceleration due to gravity. The gravitational potential $\psi_g = -gz$ (J kg^{-1}) is negative because work must be done in order to raise water from any position z in the peat column to $z = 0$.

The matric potential of peat water ψ_m (J kg^{-1}) is also negative and is a function of water content. Thermo-mechanical work must be done in order to return the peat water to the reference state, by removing it from the peat, whether by suction, drainage or evaporation.

The relationship between water content and matric potential, the peat–water characteristic $\psi_m = \psi_m(\theta)$, is a constitutive property of the peat–air–water system and is determined by measurements on peat samples at the pedon scale ($\approx$1 m). The inverse function $\theta = \theta(\psi_m)$ and its derivative, the specific water capacity $C(\psi_m) = d\theta/d\psi_m$, may be used to eliminate θ from equation (1), k_l and k_v, yielding θ_m as the dependent variable. θ may also be chosen as the dependent variable. The choice of ψ_m is preferable, since ψ_m may be a continuous function of z when θ is not. This is the case when ρ_b and k_l are discontinuous functions of depth.

Combining equations (11), (12) and (13) and eliminating the fluxes, we find, for the unknown matric potential $\psi_m(z, t)$:

$$\rho_w C \frac{\partial \psi_m}{\partial t} = \frac{\partial}{\partial z}\left[k \frac{\partial \psi_m}{\partial z} - kg\right] \qquad (14)$$

where $k = k_l + k_v$. When $k_v = 0$ this is known as Richards' equation (1931) of soil physics; the extension to include vapour transport is due to Philip (1955). This is a non-linear parabolic equation of the diffusion type with "concentration" dependent coefficients $C(\psi_m)$ and $k(\psi_m)$.

Extensions of Richards' equation to include two or three space dimensions, hysteresis, the non-rigid matrices of peats, the presence of electrolytes and the effect of temperature gradients, are also possible. The above presentation is sufficient for this discussion.

3.2. Potential energy and the phase-partitioning of water

The potential energy, ψ, of water in a state S is equal to the notional work that must be performed by a given process, in order to convert water from a reference state S_0, to the state S. The reference state is free pure liquid water at $z = 0$ in isothermal equilibrium with its vapour at saturation in a standard atmosphere.

Liquid water in an unsaturated peat is not free, since it is bound to the peat matrix by capillary and adhesive forces. Furthermore the concentration of water vapour in the peat–air, which is in equilibrium with liquid peat–water, has a value that is less than that in the reference atmosphere.

The work done, $\psi_h < 0$, in expanding isothermally the water vapour in the standard atmosphere in S_0, to achieve a lower humidity is:

$$\psi_h = \frac{RT}{M_w} \ln h \qquad 0 < h < 1 \qquad (15a)$$

assuming that atmospheric water vapour obeys the ideal gas and partial pressure laws. M_w is the molar mass of water (0.018 kg/mol), R the molar gas constant (8.314J/molK) and T the constant temperature in degrees Kelvin (293°K at 20°C).

ψ_m (< 0) is the work done in transferring isothermally the liquid water from the state S_0 into an inert rigid porous peat. We may equate ψ_m and ψ_h when the liquid and vapour phases of peat–water are in thermodynamic equilibrium. Inverting the logarithm we find the relative humidity of peat–air h as a "thermodynamic function" of peat–water potential ψ_m at any z (Edlefsen and Anderson, 1943; Penman, 1958; Rose, 1966):

$$h(z) = \exp\left[\frac{M_w \psi_m(z)}{RT}\right] \tag{15b}$$

Given $\psi_m(z,t)$, relationship (15b) instantaneously partitions $\theta(\psi_m)$ between the liquid and vapour phases, and allows us to use one, rather than two, state variables for peat–water content.

In terms of thermodynamics, ψ, is the partial specific Gibbs free energy of the water in the peat–air–water system (Edlefsen and Anderson, 1943). The work performed by the process in converting water from state S_0 to S must be reversible, independent of path, and depend solely on the initial and final states. Necessary conditions for this, are usually given as: (a) the process must be so slow that the water passes through equilibrium states only, (b) no friction, and (c) constant temperature (Rose, 1966).

Condition (a) contradicts equation (14), since this parabolic equation has the property that changes in ψ_m are propagated everywhere with infinite speed. For a penetrating critique of the problematic foundations of classical thermodynamics and its development on a rigourous foundation, see Truesdell (1981, 1984) and Serrin (1986).

3.3. Liquid and vapour conductivities

k_l and k_v are the conductivities (kg s m^{-3}) for liquid and water vapour in peat at the pedon scale. They depend on the properties of the peat matrix and on water content θ. The inverse function $\theta = \theta(\psi_m)$ of the peat–water characteristic may be used to express $k_l[\theta(\psi_m)]$ and $k_v[\theta(\psi_m)]$ as functions of ψ_m.

k_l is zero at zero water content, and increases monotonically to a maximum at saturation. k_l is determined by measurement on peat samples at the pedon scale. In contrast to k_l, k_v is zero for both saturated and completely dry peat, and may rise to a maximum at a potential of roughly -10^4 J kg^{-1} (equivalent to -10^3 m of water, pF = 3, when potential is defined as energy per unit weight). In the early stages of drying, the relative humidity of peat air is 1 and k_v is negligible compared with k_l. However, when the water potential in the peat drops below -10^4 J kg^{-1}, mass transfer takes place almost entirely in the vapour phase.

The flux density of water vapour f_v is given by the isothermal theory of vapour diffusion (Fick's law) in porous media (Penman, 1940; Philip, 1955):

$$f_v = -D_v \frac{dc}{dz} \tag{16a}$$

where c is the water vapour concentration in peat air (kg m^{-3}) and D_v the corresponding vapour diffusivity ($m^2 \, s^{-1}$). D_v is given by the expression:

$$D_v = -\alpha D_0[\theta_s - \theta(\psi_m)] \tag{16b}$$

where D_0 is the binary diffusion coefficient of water vapour in still air, at the prescribed temperature; $[\theta_s - \theta(\psi_m)]$ is the air-filled porosity of tortuosity α, through which vapour diffusion can occur.

Accordingly the conductivity k_v is:

$$k_v = -\alpha D_0(\theta_s - \theta)c_{sat}\frac{\partial h}{\partial \psi_m} \tag{16c}$$

where $h = c/c_{sat}$ is the relative humidity and c_{sat} is the value of c at saturation for the given temperature T. From expression (15b) we can find $dh/d\psi_m$. Hence the vapour conductivity k_v can be expressed entirely in terms of ψ_m.

In a shallow layer of milled peat, the air entrained within it, may not be entirely still. D_0 may be multiplied by a factor to allow for this.

4. Boundary conditions at the pedon scale

4.1. The drainage boundary condition at the bottom

The boundary condition at the bottom ($z = z_B$) of the column of unsaturated peat can be expressed in terms of the air-entry potential ψ_e

$$\psi = \psi_e \qquad \text{at } z = z_B \tag{17a}$$

$\psi_e < 0$ is the potential at which air first enters a saturated ($\theta = \theta_s$) sample of peat during drying, coinciding with the emptying of the largest pores in the peat matrix. This condition fixes the top of the saturated capillary fringe at a depth z_B below the surface. We will refer to z_B as the depth to the water table.

Water fluxes in either direction are possible across the lower boundary at z_B and represent capillary rise from or drainage to the saturated zone in the lower deposits of peat. A fixed value of z_B is adequate for simulating periods of infiltration or evaporation of a few days duration.

On longer time scales, z_B varies with time. z_B increases due to outflow of water to drains, and it decreases when rain-water infiltrates down to z_B. During prolonged periods of high evaporation z_B may also decrease.

We may derive a differential equation for $z_B(t)$ from a dynamic water balance on a column of saturated peat lying between z_B and z_D, the water level in the drains.

$\phi_f \cdot (z_D - z_B)$ is the volume of water in a column of porosity ϕ_f per unit cross-sectional area. Assuming that the flow rate to (and from) the drains is proportional to the volume of water in the saturated column, we write the dynamic water balance as:

$$\rho_w \phi_f \frac{dz_B}{dt} = k \cdot \rho_w \phi_f (z_D - z_B) - f_l(z_B) \tag{17b}$$

where k (s^{-1}) is a fractional rate of adjustment in the relative water level ($z_D - z_B$) due to drainage. $f_l(z_B)$ is the flux density of liquid water (12a) at $z_B(t)$. We may ignore the vapour flux contribution in this case. Equation (17b) is a first-order inhomogeneous ordinary differential equation for $z_B(t)$. It is also linear when the porosity ϕ_f is independent of depth. $f_l(z_B)$ couples equations (14) and (17b).

Equation (14) must now be solved subject to a moving boundary condition (17b). This is considerably more difficult than the fixed condition in (17a). The parameter k may be found using field measurements of $z_B(t)$ or simple models of saturated flow to drains. A more detailed treatment requires a two-dimensional theory of saturated/unsaturated flow.

4.2. Surface boundary condition for infiltration of rain

The boundary condition for infiltration of rain into peat is a prescribed flux at $z = 0$, equal to the rainfall rate Q ($m^3\ m^{-2}\ s^{-1}$):

$$f_l = -k_l \frac{\partial \psi}{\partial z} = \rho_w Q \qquad \theta < \theta_s \text{ at } z = 0 \tag{18}$$

Vapour flux is not normally considered during infiltration.

As the solution of equations (14), (17a) and (18) evolves in time, from a given initial condition $\theta(z, 0)$, $\theta(0, t)$ increases and may reach saturation. When saturation occurs, the first phase of infiltration at the atmosphere-controlled rate is completed and the boundary condition switches to a prescribed concentration $\theta = \theta_s$:

$$\theta(0, t) = \theta_s \tag{19}$$

As the solution of equations (14), (17a) and (19) evolves in time, the flux into the peat $f_l(0, t)$ decreases from its initial value of Q. During the second phase of infiltration, the surplus water accumulates on the surface of the ground and flows horizontally to surface drains.

In the first phase, the flux into the peat is atmosphere- controlled. In the second phase, it is peat-controlled. Analytical solutions for both linear and non-linear simplifications of equations (14), (17a), (18) and (19) are summarised in Kühnel (1989). Approximate expressions for the "time to ponding" for soils and consequently peats as well, that are initially dry, are given in Kühnel et al. (1990a). Time to ponding may vary by several orders of magnitude depending on peat properties and rainfall. If the rainfall intensity is less than the liquid conductivity at saturation, the switch to the second phase does not occur.

4.3. Surface boundary condition for isothermal evaporation

We now consider the peat after rainfall has stopped. The patches of ponded water, if there are any, will gradually disappear due to infiltration and evaporation. The infiltrated water will also be redistributed within the peat due to matric forces and gravity.

In order to simplify the presentation, we begin with a column of bare peat with an initial water content but no ponded water on the surface. The initial water content is the accumulated effect of past rainfall, evaporation and drainage, modified by spatially varying topography, atmospheric and peat properties. The relative humidity of the air in the peat surface is 1 and evaporation begins at the potential rate E_p.

The boundary condition for evaporation at the surface is a flux E of water vapour from peat to atmosphere:

$$f_l = -k_l \frac{\partial \psi}{\partial z} = 0 \qquad \text{at } z = 0 \tag{20a}$$

$$f_v = -k_v \frac{\partial \psi}{\partial z} = -\rho_w E \qquad \text{at } z = 0 \tag{20b}$$

The actual evaporation rate E (m s^{-1}) may be defined as follows (Brutsaert, 1982):

$$E = H(w)[c_0 - c_a] \tag{21}$$

where $H(w)$ is a mass transfer coefficient that varies with wind speed w at a reference height above the ground. c_0 is the water vapour concentration in the air in the peat surface and c_a is the water vapour concentration at the reference height in the atmosphere. In order to combine this boundary condition with the Richards/Philip equation, we must relate c_0 to ψ_m at $z = 0$.

The simplest relationship between c_0 and $\psi_m(0)$ is obtained by assuming an isothermal peat–atmosphere continuum. Hence we may define a potential or maximum evaporation rate E_p (m s^{-1}) as the evaporation rate from a wet peat at the same uniform temperature as the atmosphere:

$$E_p = H(w)[c_{sat} - c_a] \tag{22}$$

where c_{sat} is the saturation water vapour concentration in the peat surface at the given temperature.

Numerical experiments by van Bavel and Hillel (1976), based on an albedo that varies with water content, show that potential evaporation is not a precise concept even under isothermal conditions. In general, evaporation depends on both the peat and the atmosphere. Nevertheless we shall find it a very useful first approximation.

Eliminating $H(w)$ from (21) and (22) we find (Philip, 1957a):

$$E = E_p \left[\frac{h_0 - h_a}{1 - h_a}\right] \qquad 1 \geq h_0 \geq h_a \tag{23}$$

where $h_a = c_a/c_{sat}$ is the relative humidity of the atmosphere at the reference height; $h_0 = c_0/c_{sat}$ is the relative humidity of the air in the peat surface at $z = 0$, and is found by inserting $\psi_m(0)$ into (15b):

$$h_0 = \exp\left[\frac{M_w \psi_m(0)}{RT}\right] \tag{24}$$

During the isothermal drying of wet peats, the precise value of h_a is not important in equation (23) for small values of t, since $h_0 = 1$, $E/E_p = 1$ and E_p drives the model. Eventually h_0 may decrease from 1 at a rate that depends on peat hydraulic properties which control the upward supply of water. The evaporating surface, which is initially located in the peat surface at $z = 0$, starts to move downwards and transport of peat–water in the vapour phase appears. Eventually thermal gradients become important and must be considered.

Values of E_p may be estimated from Penman's (1948) equation using standard meteorological data or from open water evaporation rates after correction by a pan factor. A diurnal cycle may be introduced into E_p in proportion to the elevation of the sun (Van Bavel and Hillel, 1976; Hillel, 1977).

Combining equations (20b), (23) and (24), we find a non-linear inhomogeneous boundary condition of the mixed type at $z = 0$:

$$k_v \frac{\partial \psi_m}{\partial z} - \frac{\rho_w E_p}{1 - h_a} \exp\left[\frac{M_w \psi_m}{RT}\right] = k_v g - \frac{\rho_w E_p h_a}{1 - h_a} \tag{25}$$

for isothermal evaporation from wet bare peat.

5. The drying of a layer of milled peat

5.1. Modelling milling and harrowing

In the model presented in the previous sections, peat properties may vary with depth. We can model the action of milling as a modification of the physical properties of peat over the appropriate depth. These changes must be determined by laboratory and field measurements at the pedon scale. In addition both milling and harrowing may be assumed to redistribute the peat water uniformly over the milled layer.

We may envisage at least two layers distinguished by different properties. The top layer is the milled layer of thickness z_M sitting on top of a layer or column of drained peat extending to the water table at z_B. We may consider the peat matrix in both layers as rigid and having different dry bulk densities ρ_b, total fluid porosities ϕ_f, saturation volume wetnesses θ_s, peat–water characteristics $\psi_m(\theta)$ and liquid conductivity functions $k_l[\theta(\psi_m)]$.

In order to reduce the dimensionality of the parameter space, the following parametric expressions for ψ_m, C and k_l may be used (Brooks and Corey, 1966):

$$\psi_m = \psi_e \left(\frac{\theta}{\theta_s}\right)^{-b} \qquad \psi_m \leq \psi_e < 0 \tag{26a}$$

$$\theta = \theta_s \left(\frac{\psi_e}{\psi_m}\right)^{1/b} \qquad 0 \leq \theta \leq \theta_s < 1 \tag{26b}$$

$$C = \frac{d\theta}{d\psi_m} = -\frac{\theta}{(b\psi_m)} \tag{27}$$

$$k_l = k_s \left(\frac{\theta}{\theta_s}\right)^{m} \qquad m = 2b + 2 + p \tag{28a}$$

$$k_l = k_s \left(\frac{\psi_e}{\psi_m}\right)^{n} \qquad n = \frac{(2b + 2 + p)}{b} \tag{28b}$$

These relationships depend on the five parameters ψ_e, θ_s, k_s, b and p. The parameter ψ_e is the potential at which air first enters a saturated ($\theta = \theta_s$) sample of peat during drying, coinciding with the emptying of the largest pores in the peat matrix. Values of ψ_e from -1 to -2.5 J/kg have been reported for some Irish peat soils (Brandyk et al., 1985; Galvin, 1976).

The parameter b is the percentage increase in ψ_m/ψ_e for a 1% decrease in the saturation ratio θ/θ_s. When $b = 0$ all the water in the peat is held at a single potential. As b approaches infinity, no change in water content occurs when ψ_m is changed. Values of b between 3 and 15 have been reported for some Irish peat soils (Brandyk et al., 1985; Galvin, 1976).

The parameter k_s is the hydraulic conductivity at saturation θ_s. Values of k_s from 0.00025 kg s m^{-3} to 0.000005 kg s m^{-3} have been reported for some Irish peat soils (Galvin, 1976), a range of 200. Since the largest pores are believed to empty first, k_s declines rapidly to zero for values of θ less than saturation.

The expressions for ψ_m and k_l share the parameter b in accordance with the theory of flow through pores of different size, that are randomly distributed throughout a rigid porous medium (Brooks and Corey, 1966). The parameter p is a pore interaction parameter which facilitates the fitting of laboratory and field data. p has a value of 1 for some agricultural soils (Campbell, 1974).

Milling decreases the bulk density of peat and consequently increases the volumetric saturation θ_s, which occurs in expressions (26) and (28). Because of the uncertainty in the value of this and other changes due to milling and harrowing, the assumption of a rigid matrix is a reasonable starting point.

5.2. *Numerical treatment*

Campbell's (1985) integrated finite difference scheme is used. The range of z is divided into a set of elements separated by nodes. The ith element runs from the node at z_i to the node at z_{i+1}. The flux f_i through element $_i$ due to the difference in matric potential across the element is approximated by:

$$f_i = -\frac{[k_{i+1}\psi_{i+1} - k_i\psi_i]}{(1-n)[z_{i+1} - z_i]} \tag{29}$$

where ψ_i and k_i, and ψ_{i+1} and k_{i+1}, are the matric potentials and conductivities at nodes i and $i + 1$ respectively. The flux U_i at node i into element i due to the gravitational potential is:

$$U_i = gk_i \tag{30}$$

The isothermal vapour flux $f_{v(i)}$ in each element i is approximated by:

$$f_{v(i)} = D_v c_{sat} \frac{[h_{i+1} - h_i]}{[z_{i+1} - z_i]} \tag{31}$$

where $h_i = c_i/c_{sat}$ is the relative humidity at node i. D_v is given by expression (16b).

We may now write the mass balance for node i: $F_i = 0$, as follows:

$$f_{i-1} + U_{i-1} + f_{v(i-1)} - f_i - U_i - f_{v(i)} - \rho_w \frac{[\theta_i^{j+1} - \theta_i^j][z_{i+1} - z_{i-1}]}{2\Delta t} = 0 \tag{32}$$

The first three terms are the flux densities of peat water (kg m^{-2} s^{-1}) into node i. The second three terms are the flux densities out of node i. The last term is the storage of water at node i approximated using a centred difference in space and a forward difference in time. The index j corresponds to points in time which are typically 1 hour apart. The boundary conditions are entered in the first and last equations.

The numerical problem is to find the values of the potentials ψ_i at each node and at each point in time j so that the mass balance is always satisfied: $F_i = 0$. The equations form a non-linear set of tri-diagonal equations in the unknown potentials at the forward time-step. A Newton-Raphson iterative technique may be used to solve them. For further comments on the scheme, see Campbell (1985). For favourable comments on its application and on the limitations of the parametric relationships used, see Scotter et al. (1988). Vauclin et al. (1979) provide a comparative survey of other numerical techniques for solving Richards' equation.

The results in Figures 1 and 2 were calculated using this scheme.

5.3. The two phases of evaporation

Philip (1957b, 1967) concluded for isothermal steady-state evaporation from water tables at various depths, that the rate of evaporation is virtually independent of the surface relative humidity — and therefore of water content also — in the range $0 \leq h_0 \leq 0.99$. The reason is that the peat–water diffusivity function $D = k/C = k(\psi_m) \cdot d\psi_m/d\theta$ decreases through several orders of magnitude as relative humidity declines from 1 by as little as 1%. This corresponds to a downward movement of the evaporating surface and the emergence of a vapour bottleneck in the top of the peat. Other papers that discuss this very important feature of evaporation from soils are Rose (1963), Jackson (1964) and Sasamori (1970).

Once the relative humidity of the peat surface (h_0) has decreased below about 0.99, the dominant influence on the evaporation process is the resistance to water movement in the peat and not conditions in the atmosphere. Consequently we may distinguish between two phases of evaporation:

(1) atmosphere-controlled or potential evaporation when $h_0 = 1$, $E/E_p = 1$;

(2) peat-controlled evaporation when $1 > h_0 > h_a$, $E/E_p < 1$.

Under constant potential evaporation E_p, the switching from phase 1 to phase 2 evaporation, when it occurs, is sudden and precipitate at the pedon scale. The ratio of actual to potential evaporation E/E_p falls rapidly during several hours. This can be verified by numerical solution of equation (14) subject to boundary conditions (17a) and (25). These features have also been observed in many laboratory experiments on soils and references may be found in Philip (1957b). They have also been observed during experiments on milled peat samples in a tunnel dryer (S. Ward, pers. commun.). Figures 1 and 2 illustrate these phenomena.

Consequently, we may imitate the switched boundary conditions for infiltration, and replace the single mixed non-linear boundary condition (25) with: (a) a prescribed evaporation flux E_p during the first phase of evaporation, and (b) a prescribed surface potential $\psi_m(0,t)$ during the second phase.

The switching condition for evaporation may be set as a relative humidity of 0.99 or as the corresponding potential or water content at $z = 0$. Peats with a shallow water table and sufficiently small values of E_p may never enter the second phase of drying.

Analytical solutions for both linear and non-linear simplifications of equations (14) and (17a), and these switched boundary conditions are summarised in Kühnel (1989). Approximate expressions for the time to switching for soils, including peat soils, that are initially saturated, are given in Kühnel et al. (1990b). This time may vary by orders of magnitude depending on soil properties and potential evaporation.

The concepts of the atmosphere-determined and soil-determined phases of isothermal drying, are good approximations, but are not exact. Simulations by Sasamori (1970) of coupled heat, momentum and water flux in a combined soil–atmosphere model, show that the approximations remain good when a more complete treatment of the physics is considered.

Philip (1975) has remarked that "evaporation from an initially wet swelling soil will not exhibit the sharp transition between constant-rate and falling-rate phases which is characteristic of non-swelling soils". Peat soils may also crack as they dry. These phenomena are not included in the model presented here, but they may be less important in milled peat that has been shredded.

The influence of thermal gradients on water flux becomes important as the second phase of evaporation proceeds. This leads to a non-isothermal theory of evaporation, the energy and radiation balance of the surface of the ground, and heat, mass and momentum transfer in the lower boundary layer of the atmosphere (Philip and de Vries, 1957). For more recent work, see Passerat de Silans (1986).

5.4. The role of harrowing

The function of harrowing in the production of milled peat is to return the peat surface to the first phase of evaporation, in order to capture as much wind and solar energy as possible. The second phase of drying must be present at the end of the production cycle, if a low final water content is to be achieved. This depends strongly on the depth of the milled layer and on the availability of solar and wind energy.

A peat-drying simulator has been programmed for a PC-AT which allows the user to examine, graphically and numerically, the interacting effects of the following control variables: (1) the depth to the drained water table, (2) the depth of the milled layer, and (3) the timing and frequency of harrowing. The numerical scheme described in the previous sections is the basis of the program. We may use the simulator to calculate the answers to certain questions concerning the timing and frequency of harrowing, given a production interval — 4 days in our case, a target volume wetness, peat properties and a potential evaporation rate. Figures 1 and 2 show the effect of harrowing and depth of milling on evaporation and final water content. We take an average potential rate of evaporation of 3 mm/day.

Figure 2A shows the variation in the efficiency of evaporation from a milled layer initially 40 mm thick. This efficiency is defined as the ratio of actual evaporation E to potential evaporation E_p. The four-day simulation starts from uniformly wet conditions in a column of peat, standing 1 meter above the water table. Typical peat properties are used, as indicated earlier. Milling is assumed to change these values in a plausible way, increasing the saturated hydraulic conductivity and reducing the air-entry potential and the exponent b in the peat–water characteristic relationship. No data exist to guide this choice at present.

Roughly 30 hours after the start of the simulation, the evaporative efficiency suddenly falls from 1, marking the onset of the second phase of evaporation or peat-limited drying. The gradual decrease in the average volumetric water content of the milled layer is shown as curve A in Figure 2C. It terminates after 4 days at a water content of 63%. Figure 2B shows the response in evaporative efficiency when the spread of milled peat is harrowed at 60, 70, 80 and 88 hours after the start of drying. Curve B in Figure 2C shows the effect of this on the average volumetric water content of the milled layer. The final value is 54%, a decrease of 9% on 63% indicating the importance of harrowing. The function of harrowing is to increase the area under the efficiency curve by returning the evaporation from peat-control to atmosphere-control after the second phase of drying has started.

Figure 1 compares two different regimes of harrowing in the case of a layer 50 mm thick. In Figure 1B, harrowing is applied at intervals of 18 hours, i.e. at 18, 36, 54 and 72 hours. The first two harrowings cannot be seen in Figure 1B. The

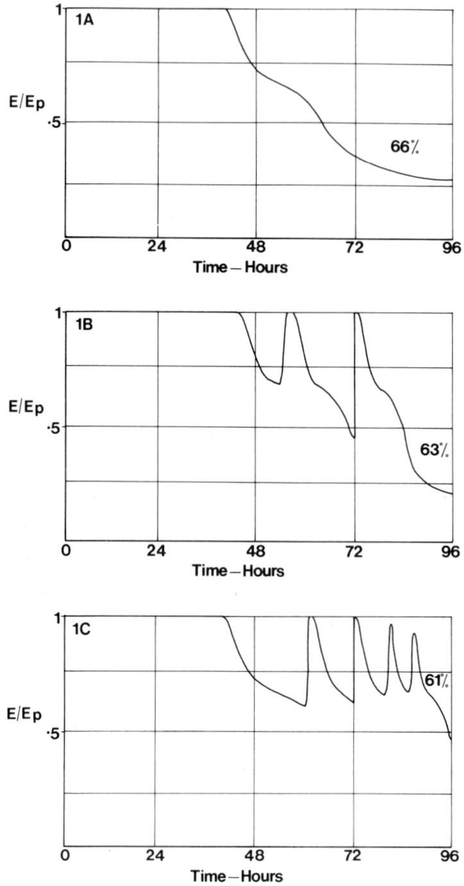

Fig. 1. Case 1A. The variation in evaporative efficiency over four days in the case of a 50 mm layer of milled peat. The final volumetric water content is 66%. Case 1B. The variation in evaporative efficiency, when the 50 mm layer is harrowed at 18, 36, 54 and 72 hours after the start of drying. The final volumetric water content is 63%. Case 1C. The variation in evaporative efficiency, when the 50 mm layer is harrowed later than in 1B, i.e. at 60, 72, 80 and 86 hours after the start of drying. The final volumetric water content is 61%.

final water content of the milled layer is 63%, which is the same as the case shown in Figure 2A, i.e. no harrowing applied to a 40 mm layer. Clearly, harrowing can compensate for an increase in the thickness of the milled layer.

Figure 1C examines a later application of harrowing, at 60, 72, 80 and 86 hours. In this case, all four harrowings are visible in the evaporative efficiency. They yield a final water content of 61%. We conclude that the timing of harrowing is also important. The results also suggest that: (a) harrowing should start very soon after the onset of the second phase of drying, and (b) the subsequent intervals between harrowings should decrease with time as the milled layer dries

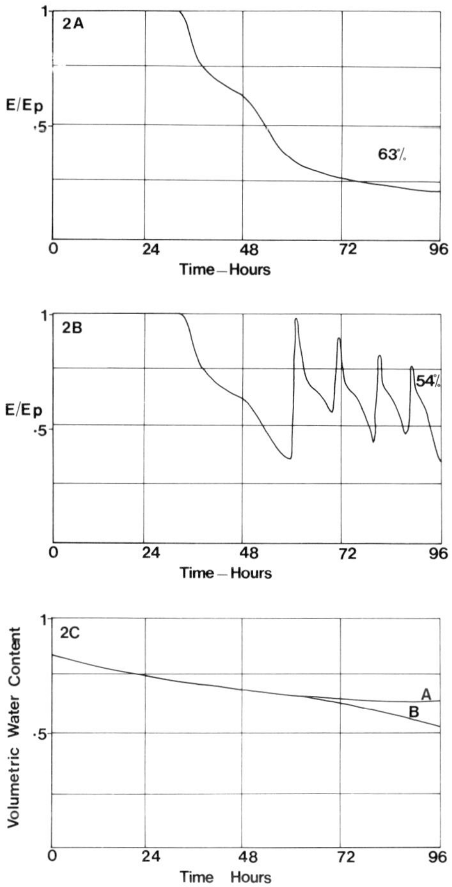

Fig. 2. Case 2A. The variation in evaporative efficiency over four days in the case of a 40 mm layer of milled peat. The final volumetric water content is 63%. Case 2B. The variation in evaporative efficiency, when the 40 mm layer is harrowed at 60, 70, 80 and 88 hours after the start of drying. The final volumetric water content is 54%. Case 2C. The variation in volumetric water content in a 40 mm layer, with no harrowing (*A*) and with harrowing applied at 60, 70, 80 and 88 hours after the start of drying (*B*).

out. These suggestions are consistent with a possible overall objective of maximising the area under the efficiency curve, after the second phase of drying has started.

6. Conclusions

The hydrology of milled peat production contains a number of interesting questions on the effectiveness of the major control variables in promoting the evaporation of water: (1) the depth to the drained water table, (2) the depth of the milled layer, and (3) the timing and frequency of harrowing.

Tentative answers have been presented in the last section. These are: (1) for a given target water content, harrowing can compensate for an increase in the depth of the milled layer, (2) the timing of harrowing has a measurable impact on the final water content, and (3) harrowing might be applied to maximise the area under the evaporative efficiency curve.

Further progress requires a matching programme of laboratory and field experimentation, and the extension of the model to include the diurnal cycle, the energy balance at the surface, thermal gradients and a non-rigid matrix. The spatial variability of peat properties is an additional complication.

Acknowledgements

Aspects of this research have been supported by Bord na Móna (Peat Development Board) and by the Climatology Research Programme of the European Communities under the projects "The role of soil water in global climate modelling" and "Spatial variability of land surface processes". It also received an International Todd Andrews award.

I wish to acknowledge the assistance of John Turner in completely rewriting the original peat-drying simulator, and my former colleagues in CWRR, James Dooge, Vit Kühnel and Graham Sander, for many valuable discussions. Interest in the problems of peat production was stimulated by my former colleague Shane Ward of the Department of Agricultural and Food Engineering, UCD.

References

Bellamy, D., 1986. The *Wild* Boglands — Bellamy's Ireland. Wild Ireland Library. Country House, Dublin, 178 pp.

Brandyk, T., Dodd, V.A. and Grace, P.M., 1985. As assessment of some unsaturated hydraulic conductivity characteristics for a range of Irish peat soils. Ir. J. Agric. Res., 24: 237–251.

Brooks, R.H. and Corey, A.T., 1966. Properties of porous media affecting fluid flow. J. Irrig. Drain., Div. Am. Soc. Civ. Engineers, 72, IR-2, pp. 61–88.

Brutsaert, W., 1982. Evaporation into the Atmosphere. D. Reidel, Dordrecht, 299 pp.

Buckingham, E., 1907. Studies on the movement of soil moisture. U.S. Dep. Agric. Bur. Soils Bul., 38: 28–31.

Campbell, G.S., 1974. A simple method for determining unsaturated conductivity from moisture retention data. Soil Sci., 117: 311–314.

Campbell, G.S., 1977. An Introduction to Environmental Biophysics. Springer Verlag, New York, N.Y.

Campbell, G.S., 1985. Soil Physics with BASIC — Transport Models for Soil–Plant Systems. Elsevier, Amsterdam, 150 pp.

Dooge, J.C.I., 1975. The water balance of bogs and fens. Proc. Minsk Symp. on the Hydrology of Marsh-Ridden Areas. IHD Studies and Reports on Hydrology, UNESCO, Paris, 19: 233–272.

Edlefsen, N.E. and Anderson, A.B.C., 1943. The thermodynamics of soil moisture. Hilgardia, 16: 31–299.

Galvin, L.F., 1976. Physical properties of Irish peats. Ir. J. Agric. Res., 15: 207–221.

Hillel, D., 1977. Computer Simulation of Soil–Water Dynamics; A Compendium of Recent Work. International Development Research Centre, Ottawa, p. 214.

Jackson, R.D., 1964. Water vapour diffusion in relatively dry soil: I. Theoretical considerations and sorption experiments. Soil Sci., Soc. Am. Proc., 28: 172–176.

Kühnel, V., 1989. Scale problems in soil moisture flow. Ph.D. Thesis. National University of Ireland.

Kühnel, V., Dooge, J.C.I., Sander, G.C. and O'Kane, J.P., 1990a. Duration of atmosphere-controlled and of soil-controlled phases of infiltration for constant rainfall at a soil surface. Ann. Geophys., 8(1): 11–20.

Kühnel, V. Dooge, J.C.I., Sander, G.C. and O'Kane, J.P., 1990b. Duration of atmosphere-controlled and of soil-controlled phases of evaporation for constant potential evaporation at a soil surface. Ann. Geophys., 8(1): 21–28.

O'Kane, J.P., Kühnel, V. and Sander, G., 1987. Mathematical Simulations of Drying Peat — Final Report for Bord na Móna. Centre for Water Resources Research, Dept. Civ. Eng., University College Dublin.

Passerat de Silans, A., 1986. Transferts de masse et de chaleur dans un sol stratifié soumis à une excitation atmosphérique naturelle. Comparison: modèles–expérience. Thèse de Docteur de l'Institut National Polytechnique de Grenoble.

Penman, H.L., 1940. Gas and vapour movement in soil. I and II. J. Agric. Sci., 30: 437–462 and 570–581.

Penman, H.L., 1948. Natural evaporation from open water, bare soil and grass. Proc. R. Soc. London, Ser. A, 193: 120–145.

Penman, H.L., 1958. Humidity. Monograph of the Institute of Physics, Chapman and Hall, London.

Philip, J.R., 1955. The concept of diffusion applied to soil water. Proc. Natl. Acad. Sci. India, Sect. A, 24: 93–104.

Philip, J.R., 1957a. The physical properties of soil water movement during the irrigation cycle. Proc. 3rd. Int. Congr., Int. Comm. on Irrigation and Drainage, pp. 8.125–8.154.

Philip, J.R., 1957b. Evaporation, and moisture and heat fields in the soil. J. Meteorol., 14: 354–366.

Philip, J.R., 1967. The second stage of drying of soil. J. Appl. Meteorol., 6: 581–582.

Philip, J.R., 1975. Water movement in soil. In: D.A. de Vries and N.H. Afgan (Editors), Heat and Mass Transfer in the Biosphere. Papers from a Forum on Transfer Processes in the Plant Environment, held in Dubrovnik, August 1974. Adv. Therm. Eng., 3: 29–47.

Philip, J.R. and de Vries, D.A., 1957. Moisture movement in porour materials under temperature gradients. Trans. Am. Geophys. Union., 38: 222–232.

Richards, L.A., 1931. Capillary conduction of liquids through porous mediums. J. Phys., Nov., 1: 318–333.

Rose, C.W., 1966. Agricultural Physics. Pergamon Press, London.

Rose, D.A., 1963. Water movement in porous materials. Part 1 — Isothermal vapour transfer. Br. J. Appl. Phys., 14: 256–262.

Sasamori, T., 1970. A numerical study of atmospheric and soil boundary layers. J. Atmos. Sci., 27: 1122–1137.

Scotter, D.R., Clothier, B.E. and Sauer, T.J., 1988. A critical assessment of the role of measured hydraulic properties in the simulation of absorption, infiltration and redistribution of soil water. Agric. Water Manage., 15: 73–86.

Serrin, J. et al., 1986. New Perspectives in Thermodynamics. Springer-Verlag, Berlin.

Truesdell, C., 1981. Textbook of Classical Thermodynamics. Department of Mechanics, The Johns Hopkins University.

Truesdell, C. et al., 1984. Rational Thermodynamics. 2nd Edition. Springer-Verlag, Berlin.
Van Bavel, C.H.M. and Hillel, D.I., 1976. Calculating potential and actual evaporation from a bare soil surface by simulation of concurrent flow of water and heat. Agric. Meteorol., 17: 453–476.
Vauclin, M., Haverkamp, R. and Vachaud, G., 1979. Résolution numérique d'une équation de diffusion non linéaire. Application à l'infiltration de l'eau dans les sols non saturés. Presses Universitaires de Grenoble, 183 pp.

Chapter 4

Role of active–passive scalar relationships in evaporation from vegetated surfaces

GORDON A. McBEAN

Abstract

Evaporated water is carried away from the surface by the turbulent wind and the efficiency of that transfer is determined by the correlation between fluctuations of the turbulent vertical wind and moisture (or water vapour). An active scalar is defined as one whose variations significantly affect the density and hence buoyancy of an air parcel. Active scalars will be preferentially transferred vertically compared to passive scalars. Generally moisture will be a passive scalar and its transfer efficiency will depend on its correlation with the active (temperature) scalar. These relationships will introduce errors into the estimation of evaporation from gradients of temperature and humidity.

1. Introduction

Evaporation from vegetated surfaces plays a major role in the global water cycle, in the natural and essential processes of plants, and in modifying weather and climate (Shuttleworth, 1991). Knowledge of evaporation is also important to attempts to optimize the utilization of water in agricultural irrigation systems. The evaporation process is complex, involving plant physiology, available energy and water and turbulent motions of the wind, and its study has necessitated bringing together specialists and ideas from each area. Although the amount of water that can be lost from a surface is limited by available energy and water, subsequent water vapour or moisture transfer or flux away from the surface is controlled primarily by atmospheric turbulence; this may present another limitation to the evaporation rate. Exploration of the role of some turbulent processes in evaporation or vertical moisture flux is the subject of this paper.

Above the surface, the moisture flux will be carried by turbulent eddies of the wind. It can be shown (e.g., Wyngaard, 1991) that the change of the vertical flux with height will be much smaller than the measurement error over the lowest few tens of metres of the atmospheric boundary layer. In this "constant flux" or atmospheric surface layer, vertical fluxes of momentum, heat and moisture will be assumed to be constant with respect to height. The evaporation, E, will be assumed to be equal to the eddy moisture flux at a few metres above the surface or above the canopy, in the constant flux layer, so:

$$E = \langle w\rho_v \rangle \approx \rho_o \langle wq \rangle \qquad (1)$$

where w is the vertical component of the turbulent wind; ρ_v is the moisture or water vapour density (kg(H_2O) m^{-3}); ρ_o is the reference or average air density; q is the specific humidity (kg(H_2O) [kg(dry air)]$^{-1}$); and $\langle\,\rangle$ is the mean or average value.

The average value should be over an ensemble of events but is usually replaced by an average over time; although this has its limitations (Wyngaard, 1991), it will be assumed here that this is acceptable. It is further assumed that $\langle w \rangle = 0$ (see Wyngaard, 1991, for further discussion).

Generally, evaporation is not measured directly through (1) but is estimated through measurements of the gradient of humidity and temperature. Some assumptions are needed to relate the gradient to the flux or surface evaporation. Assuming the flux-gradient relationships are valid, then:

$$E = -\rho_o K_E \left[\frac{\partial Q}{\partial z}\right] \quad \text{and} \quad H = -\rho_o c_p K_H \left[\frac{\partial \Theta}{\partial z}\right] \qquad (2)$$

where z is height above the surface; $H = \rho_o c_p \langle w\theta \rangle$ is the sensible heat flux; c_p is the specific heat of air at constant pressure; θ is the deviation of the potential temperature from its mean value; K_E and K_H are the turbulent diffusivities for moisture and heat (defined by these equations). Q and Θ are the mean specific humidity and potential temperature and z is the vertical coordinate. The Bowen ratio (β) is usually approximated as:

$$\beta = \frac{H}{\lambda E} \approx \left[\frac{c_p}{\lambda}\right]\left[\frac{K_H}{K_E}\right]\left[\frac{\Delta\Theta}{\Delta Q}\right] \qquad (3)$$

where λ = latent heat of condensation and assuming the finite differences (Δ) can approximate the differentials. It is normally assumed that $\alpha = K_H/K_E = 1$, so that β can be derived directly from profile measurements. In the following sections, the validity of the $\alpha = 1$ assumption will be discussed in terms of the relative transfer efficiencies for heat and moisture. Then the concept of active and passive scalars will be introduced. Finally, the impact on estimations of evaporation will be presented.

2. Efficiency of evaporation

Eddy moisture flux occurs due to the correlation r_{wq} of instantaneous vertical wind w and instantaneous fluctuation of moisture or specific humidity q(1). If there is no vertical motion (purely horizontal flow), then the moisture flux will only take place through molecular diffusion, a very slow process. In this situation, surface evaporation would be limited by the vertical transfer through the air since the near-surface concentration of moisture would increase due to evaporation from the surface, eventually preventing further surface evaporation. It can also be noted that the turbulent moisture flux will be zero if the correlation between w and $q(r_{wq})$ is zero, since:

$$E = \rho_o \langle wq \rangle = \rho_o \sigma_w \sigma_q r_{wq} \tag{4}$$

where σ is the standard deviation. Hence, to transfer a large amount of moisture vertically requires that there be active vertical motion which is correlated with variations in moisture content. If the correlation is small, there will be comparatively little flux compared to when the correlation is large (for the same σ's). For this reason, McBean and Miyake (1972) referred to r_{wx} (and its spectral counterpart) as the transfer efficiency of any scalar quantity x. Similarly, the transfer efficiencies for sensible heat and momentum fluxes can be defined through:

$$H = \rho_o c_p \langle w\theta \rangle = \rho_o c_p \sigma_w \sigma_q r_{w\theta} \tag{5}$$

$$\tau = -\rho_o \langle wu \rangle = -\rho_o \sigma_w \sigma_u r_{wu} \tag{6}$$

where τ is the vertical flux of horizontal momentum and u is the deviation of the horizontal longitudinal wind from its mean value. The ratio $r_{wq}/r_{w\theta}$ is a measure of the relative transfer efficiency of moisture transfer compared to heat transfer.

Assuming that the vertical moisture flux is equal to the total canopy evaporation, the questions that need to be asked (and answered) are: what determines the correlation r_{wq} and how does it compare with $r_{w\theta}$ and r_{wu}? Since most models for computing evaporation have built-in assumptions on these ratios, the question is of more than academic interest. Following above, it can be written:

$$\alpha = \frac{K_H}{K_E} = \left[\frac{r_{w\theta}}{r_{wq}}\right] \left[\frac{\sigma_\theta}{\Delta\Theta}\right] \left[\frac{\sigma_\theta}{\Delta Q}\right]^{-1} \tag{7}$$

Since the fluctuations of q and θ result from the interaction of vertical motion and mean gradients, it is reasonable to assume that:

$$\left[\frac{\sigma_q}{\Delta Q}\right] \approx \left[\frac{\sigma_\theta}{\Delta\Theta}\right] \tag{8}$$

so that:

$$\alpha = \frac{K_H}{K_E} \approx \frac{r_{w\theta}}{r_{wq}} \tag{9}$$

Hence, validity of the normal use of the Bowen ratio approach implies the assumption that the transfer efficiencies for moisture and heat are equal. Hill (1989) has argued that Monin-Obukhov similarity theory [that appropriate dimensionless variables will be functions of z/L where L is the Monin-Obukhov length (see Busch, 1973)] implies $r_{wq} = r_{w\theta} = r_{wx}$ where x is any scalar quantity. However, McBean (1992) contends that this depends on an erroneous assumption and will not generally be valid.

In their study of turbulent transfer mechanisms over a grass surface, McBean and Miyake (1972) found that the absolute value of the ratio $r_{w\theta}/r_{wu}$ was about 1.4 for near neutral conditions and increased as $3[-z/L_\theta]^{1/4}$ for unstable conditions.

However, the ratio $r_{w\theta}/r_{wq}$ was quite variable, being near unity for some cases and as high as 2 to 3 for some others. The ratio $r_{w\theta}/r_{wq}$ was found to be dependent on the correlation $r_{\theta q}$. When $r_{\theta q}$ is large, then the ratio is near unity; when $r_{\theta q}$ is small, the ratio is also small. Similar results were also found in a study over a prairie grass surface (McBean, 1973), where the values of $r_{\theta q}$ varied from 0.76 early in a day to 0.13 late in the same day. Early in the day, the soil moisture content of the surface layer was probably sufficiently high so that local hot spots had local humidity maxima associated with them. As the day progressed, the surface soil moisture content was depleted and hot spots may have had less than typical moisture, so the temperature–humidity correlation decreased.

More recently, Roth (1990) investigated turbulent transfers at 11–19 m above a suburban area. He found that the ratio $r_{w\theta}/r_{wq}$ varied considerably but the mean value decreased from near 3 for $r_{\theta q} = 0.2$ to near 1.2 for $r_{\theta q} = 0.8$. The ratio also decreased from 2–3 for near neutral situations to near 1.5 for $-z/L_{\mathrm{v}}$ greater than 1. These suburban results are consistent with the earlier results of McBean.

3. Concept of active and passive scalars

The turbulent scalar flux depends on the correlation of the vertical wind and the scalar. The equation for the instantaneous vertical wind is (following Busch, 1973):

$$\frac{\partial w}{\partial t} = -U\frac{\partial w}{\partial x} - u_j\frac{\partial w}{\partial x_j} + \left\langle u_j\frac{\partial w}{\partial x_j}\right\rangle - \left(\frac{1}{\rho}\right)\left(\frac{\partial p}{\partial z}\right) + \nu\left(\frac{\partial w}{\partial x_j \partial x_j}\right) - \left(\frac{g}{\rho_{\mathrm{o}}}\right)\rho \tag{10}$$

where the Cartesian tensor convention has been used for the turbulent covariance terms. The coordinate system (x,y,z) has the corresponding winds (u,v,w) with z and w in the vertical direction and x and u along the mean wind direction. p and ρ are the turbulent pressure and density. Lower case quantities are turbulent values and upper case, means. Means of all turbulence values are zero. The Boussinesq approximation (variations in density are ignored except when they determine the weight) has been assumed. For this discussion, the important term is the last one on the right hand side, the buoyancy term, $(g/\rho_{\mathrm{o}})\rho$. It can be seen that density variations directly cause accelerations in the vertical direction. In Busch (1973) and most other atmospheric texts, it is assumed that:

$$\rho = -\left(\frac{\rho_{\mathrm{o}}}{\Theta_{\mathrm{v}}}\right)\theta_{\mathrm{v}} \tag{11}$$

where Θ_{v} is the mean virtual potential temperature and θ_{v} is the turbulent virtual potential temperature. Virtual potential temperature is the effective temperature that accounts for temperature and other constituents of the air in the determination of its density through the ideal gas law. The turbulent virtual temperature is then (Busch, 1973, pp. 22–23):

$$\theta_{\mathrm{v}} = \theta(1 + 0.61Q) + 0.61q\Theta - 0.61\langle q\theta\rangle \tag{12}$$

Therefore, density variations and the buoyancy force in the vertical equation of motion depend on both temperature and specific humidity variations. Since vertical fluxes of a property, such as heat or moisture, depend on the covariance of vertical motions and that property, the direct role of a property in determining the vertical motion is important. An active scalar will be defined as a scalar whose variations affect the density of the fluid and hence directly contribute to the buoyancy force generating vertical motion. A passive scalar will be defined as one whose variations have no (or negligible) effect on the density and hence on vertical motion. Although it can be argued that most scalars will be active by the strict definition of the word, the definition of active can be interpreted to mean a scalar whose variations have the same (or greater) impact than a small (but measurable) temperature fluctuation (say 0.01°C).

To further clarify the role of active and passive scalars, it is useful to consider the turbulent kinetic energy (*TKE*) budget for horizontally homogenous, time-stationary conditions, which, to a first approximation, can be written:

$$\left(\frac{\partial}{\partial t}\right) TKE = 0 = -\langle wu \rangle \left[\frac{\partial U}{\partial z}\right] + \left[\frac{g}{\Theta}\right] \langle w\theta_v \rangle - \varepsilon \qquad (13)$$

The first term on the RHS is the mechanical production of turbulence due to the interaction of the momentum flux with the mean wind shear. The second term is called buoyant production and results from interaction of the vertical wind and density variations (as represented by the virtual potential temperature). The last term is the rate of dissipation of turbulent kinetic energy by viscous processes. Thus, local production of turbulence mechanically and through buoyancy is balanced by local dissipation. In terms of Monin-Obukhov theory, the turbulent energy budget can be written as:

$$\phi_{\mathrm{m}} - \frac{z}{L_{\mathrm{v}}} = \Phi_{\varepsilon} \qquad (14)$$

where $\phi_{\mathrm{m}} = (kz/u_*)[\partial U/\partial z]$, k is von Karman's constant, the non-dimensional wind shear; $u_* = -\langle uw \rangle$, the friction velocity; z/L_{v}, the Monin-Obukhov stability parameter; and Φ_{ε}, the non-dimensional dissipation rate. The Monin-Obukhov stability parameter can be written (McBean, 1971; Busch, 1973):

$$\frac{z}{L_{\mathrm{v}}} = \frac{z}{L_\theta} + \frac{z}{L_q} \approx \left(\frac{z}{L_\theta}\right)\left(1 + \frac{0.075}{\beta}\right) \qquad (15)$$

where:

$$\frac{z}{L_\theta} = -\left(\frac{kz}{u_*^3}\right)\left(\frac{g}{\Theta_{\mathrm{v}}}\right)\langle w\theta \rangle$$

$$\frac{z}{L_q} = -\left(\frac{kz}{u_*^3}\right)\left(\frac{0.61 g\Theta}{\Theta_{\mathrm{v}}}\right)\langle wq \rangle$$

and the introduction of β, the Bowen ratio, assumes that $\Theta \approx 300$°K.

The ratio of buoyant to mechanical production of turbulent kinetic energy is the flux Richardson number, $Rf = -[g/\Theta]\langle w\theta_v\rangle[\langle wu\rangle\partial U/\partial z]^{-1}$, can also be written as, $Rf = -[z/L_v][\phi_m]^{-1}$. For near-neutral conditions, $\phi_m \approx 1$, so z/L_v can be interpreted as the ratio of buoyant production to mechanical production of *TKE*, for near neutral conditions; this interpretation will be used in the following. When $\beta \geq 1$, the impact of the moisture fluctuations on buoyant production is less than 8%. However, for values of β between 0.7 to 0.3 (typical of results from HAPEX-MOBILHY; Goutorbe, 1991), the correction due to moisture fluctuations can be between 10 and 25%.

In his study of turbulence statistics, McBean (1971) showed that the statistics of humidity fluctuations (σ_q/q_*, where $q_* = -\langle wq\rangle/u_*$) scaled better with z/L_q than with z/L_v or z/L_θ, because the humidity fluctuations played a small role in determining buoyancy. In this case, humidity was an essentially passive scalar, while temperature was the dominant active scalar.

Since the transfer efficiency of a scalar property depends on its correlation with vertical velocity fluctuations, and that correlation depends on whether the scalar is active or passive, it follows that the transfer efficiency of a passive scalar property will depend on the correlation between that passive scalar and the dominant active scalar(s). The importance of the active scalar can be seen in the simplified schematic of turbulent transfer in Figure 1a. If we consider turbulent transfer as resulting from vertical velocity fluctuations acting over a mixing length l, then the instantaneous

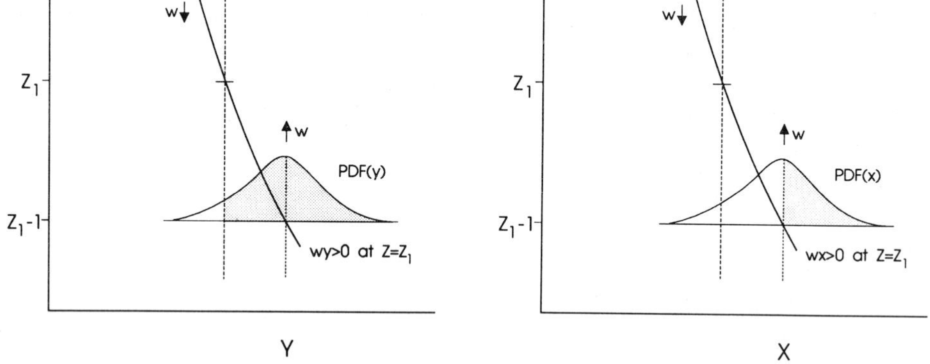

Fig. 1. Schematic of vertical transfer of a passive scalar (y, in part a of figure) and an active scalar (x, in part b of figure). The shaded region of the probability density function (*PDF*) gives the probability of positive (for $w > 0$) or negative (for $w < 0$) fluctuations of y or x. The solid heavy line is the assumed profile of the mean quantity (Y or X).

wy, where y is any passive scalar, originating at $z_1 - l$ and seen at level z_1, depends on the gradient $\partial Y/\partial z$ and the probability distribution function of y at level $z_1 - l$ [$PDF(y,z)$]. If the gradient is small and σ_y large, then wy may be almost as likely negative as positive, resulting in a low overall value of r_{wy}. If the gradient is large and σ_y relatively small, then r_{wy} will likely be large. However, if we are considering an active scalar, x, the effective *PDF* is changed since those parcels of air with $x > 0$ (at level $z_1 - l$) will be more likely to rise than those with $x < 0$. The result will be to enhance the r_{wx} value, independent of the gradient and the overall σ_x. If y is correlated with x, then it will benefit from the buoyant or active scalar role of x and increase its r_{wy}. Hence, the ratio r_{wy}/r_{wx} will depend on r_{xy} where x is an active scalar and y, a passive scalar. Similarly, the ratio K_X/K_Y will also depend on r_{xy} and it should not be assumed, a priori, that $K_X/K_Y = 1$, for all scalars. Warhaft (1976) used a model of turbulence to arrive at the same conclusion.

Mahrt (1991) has examined boundary-layer moisture regimes using data from the HAPEX-MOBILHY (André et al., 1988) and FIFE (Sellers et al., 1988). He categorizes moisture regimes in terms of the Bowen ratio and the ratio h/L, where h is the depth of the atmospheric boundary layer. In essence, this separation is in terms of the importance of moisture fluctuations in controlling buoyancy although the terms active and passive scalars are not used. A moisture/energy ratio is defined as $M = (-h/L)/\beta$, with M varying from 10–20 for windy conditions with only modest evapotranspiration to values over 100 with large evapotranspiration and weak winds. Although Mahrt focuses on the middle and upper parts of the boundary layer (since all his data is from aircraft) many of the results will also be applicable in the surface boundary layer which is also impacted by the larger eddies. It is shown that the skewness of the moisture fluctuations is related to the θ–q correlation. The w–q correlation was also found to be lower when the θ–q correlation was low.

4. Impact of K_E/K_H ratio on traditional methods of estimating evaporation from vegetated surfaces

One of the most commonly used methods of computing evaporation from vegetated (or other) surfaces is the Bowen ratio method. Since:

$$\lambda E = \frac{(R_N - G)}{(1 + \beta)} \qquad (16)$$

$$= \frac{(R_N - G)}{(1 + \alpha\beta')}$$

where $\beta' = [c_p/\lambda][\Delta\Theta/\Delta Q]$ is the usually measured approximation of the Bowen ratio. Figure 2 shows the impact of various a values on the estimation of the ratio $\lambda E = (R_N - G)$. Values of α could range from about 2 (heat transfer about twice as efficient as moisture transfer) to less than unity (moisture transfer more efficient

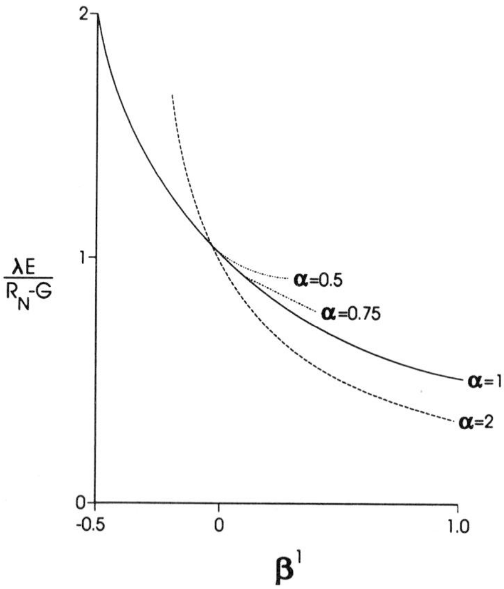

Fig. 2. Impact of a on evaporation computed by the Bowen ratio method. Computed ratio $\lambda E/(R_N - G)$ against β' for values of α = 0.5, 0.75, 1 and 2.

than heat transfer). However, the latter case is only likely for small values of β. Also, as β becomes large, the possibility of large values increases. For $\beta' = 1$ and $\alpha = 2$, the actual evaporation would only be 66% as large as the computed (assuming $\beta' = \beta$) value. Situations with $\beta' < 0$ can arise either due to negative heat flux or negative evaporative flux (condensation), but not both. Experimental observations show that the $r_{w\theta}$ correlation is small for negative heat transfer and r_{wq} can be similar or even smaller (and of either sign). The errors in computing evaporation can be relatively larger than for the $\beta' > 0$ cases and evaporation will be underestimated. For small values of β' (but $\beta' > 0$), moisture fluctuations would be the dominant active scalar and evaporation will be underestimated. For $\alpha = 0.5$, $\beta' = 0.2$, the evaporation will be underestimated by 12%.

A related method of computing evaporation is through the Penman-Monteith equation (Monteith, 1973; Rowntree, 1991). Again, it is normally assumed that $\alpha = K_H/K_E = r_a(q)/r_a(\theta) = 1$. r_a is the atmospheric resistance and the q and T refer to moisture and heat transfer resistance, respectively. Allowing for $\alpha \neq 1$, leads to:

$$\lambda E = \frac{[\Delta(R_N - G) + \rho_o c_p \delta' q/r_a(\theta)]}{[\Delta + (c_p/\lambda)(\alpha + r_s/r_a(\theta)]} \tag{17}$$

where Δ is the slope of the saturation curve with temperature, $\delta' q$ is the saturation deficit and r_s is the stomatal resistance. The only difference between (17) and the normal version is the replacement of 1 by α in the denominator. Figure 3 shows the impact of different values of α on computed $\lambda E/(R_N - G)$.

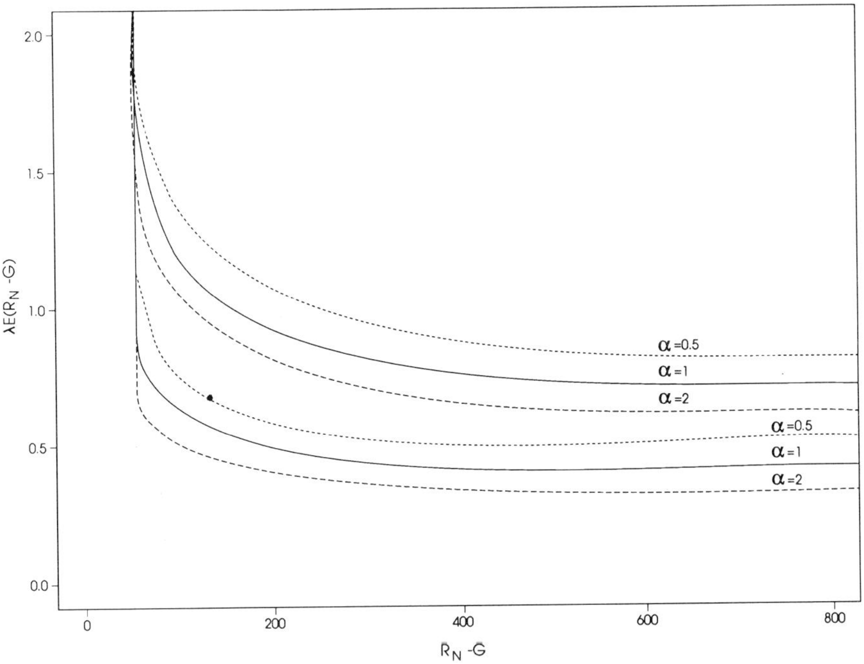

Fig. 3. Impact of α on evaporation computed by Penman-Monteith method (following Rowntree, 1991). The upper set of curves correspond to $\Theta = 30°C$, $\delta' q = 5$ g/kg and r_s/r_a ($q = 1$; the lower set), to $\Theta = 10°C$, $\delta' q = 1$ g/kg and $r_s/r_a(\theta) = 0$.

5. Concluding remarks

Vertical moisture flux results from updrafts and downdrafts of air of varying moisture content. Some of these updrafts have their origin at the surface and others, at the top of the atmospheric boundary layer. Wyngaard and Brost (1984) developed the concept of top-down and bottom-up diffusion of scalars and Mahrt (1991) has elaborated further on the concept for moisture fluctuations. Above the atmospheric boundary layer, the moisture content decreases while the potential temperature increases, so downdrafts from above will tend to be warm and dry (θ–q correlation negative). For a vegetated surface, the canopy will be a source of moisture and temperature variations. The vertical fluxes will not be constant down into the canopy layer of vegetation, where the fluxes will certainly be changing with respect to height due to local sources and sinks (e.g., Raupach, 1989). As described by Raupach, one can "regard the canopy as a collection of a vast number of scalar source elements, or tiny chimneys, puffing out plumes of scalar material" (Raupach, 1989, p. 176). As the turbulent wind interacts with these sources, some parcels will become warmer (and others relatively cooler) and hence more buoyant. Depending on the magnitude of the temperature difference, these warm parcels will ascend upward out of the vegetation into the turbulent air above. At the same time, some parcels of air will be moistened due to evapotranspiration from leaves. If the parcel is already warm, it has a larger moisture capacity and may then accumulate more moisture in its gyrations through the canopy. On the other hand, some parcels that absorb extra moisture may be neutrally buoyant and some may be relatively cool (and have negative buoyancy). The interaction of water vapour and heat sources with the turbulent wind will depend on their relative distribution through the canopy. For some canopies the sources will be essentially co-located while for others, such as tall vegetation, the upper surfaces may be warm due to intercepted solar radiation while the transpiring leaves are lower in the canopy.

When viewed from some intermediate level, above the canopy and below the top of the atmospheric boundary layer, the influences of both regions will still be seen. From this point of view, it would seem somewhat surprising if temperature and moisture fluctuations showed much correlation, but it is frequently observed that they do (Shuttleworth, 1989). The lack of correlation will cause some dissimilarity in the transfer of moisture and heat (and momentum). Since most gradient methods assume similarity of the transfer mechanisms, there will generally be an error in computing the vertical moisture flux or evaporation. However, since the turbulent correlations are only one factor in determining the flux (available energy and water being two others), the errors introduced are not overwhelming, but can be over 30%.

References

André, J.C. et al., 1988. HAPEX-MOBILHY: First results from the special observing period. Ann. Geophys., Ser. B, 6: 477–492.

Busch, N.E., 1973. On the mechanics of atmospheric turbulence. In: D.A. Haugen (Editor), Workshop on Micrometeorology. Am. Meteorol. Soc., Boston, pp. 1–67.

Goutorbe, J.P., 1991. A critical assessment of the Samer network accuracy. In: T.J. Schmugge and J.C. André (Editors), Land Surface Evaporation, Springer-Verlag, New York, N.Y., pp. 171–182.

Hill, R.J., 1989. Implications of Monin-Obukhov similarity theory for scalar quantities. J. Atmos. Sci., 46: 2236–2244.

Mahrt, L., 1991. Boundary–layer moisture regimes. Q.J.R. Meteorol. Soc., 117: 151–176.

McBean, G.A., 1971. The variations of the statistics of wind, temperature and humidity fluctuations with stability. Boundary–Layer Meteorol., 1: 438–457.

McBean, G.A., 1973. Comparison of turbulent transfer processes near the surface. Boundary–Layer Meteorol., 4: 265–274.

McBean, G.A., 1992. Monin-Obukhov similarity theory for passive scalars. J. Atmos. Sci. (submitted).

McBean, G.A. and Miyake, M., 1972. Turbulent transfer mechanisms in the atmospheric surface layer. Q.J. R. Meteorol. Soc., 98: 383–398.

Monteith, J.L., 1973. Principles of Environmental Physics. Arnold, London.

Raupach, M.R., 1989. Stand overstorey processes. Phil. Trans. R. Soc. Lond., Ser. B, 324: 175–190.

Roth, M., 1990. Turbulent transfer characteristics over a suburban surface. Ph.D. thesis, Dep. Geogr., University of British Columbia, Vancouver, 292 pp.

Rowntree, P.R., 1991. Atmospheric parameterization schemes for evaporation over land: basic concepts and climate modeling aspects. In T.J. Schmugge and J.C. André (Editors), Land Surface Evaporation Springer-Verlag, New York, N.Y., pp. 5–30.

Sellers, P.J., Hall, F.G., Asrar, G., Strebel D.E. and Murphy, R.E., 1988. The first ISLSCP Field Experiment (FIFE). Bull. Am. Meteorol. Soc., 69: 22–27.

Shuttleworth, W.J., 1989. Micrometeorology of temperate and tropical forest. Phil. Trans. R. Soc. Lond., Ser. B, 324: 299–334.

Shuttleworth, W.J., 1991. The modellion concept. Rev. Geophys., 29: 585–606.

Warhaft, Z., 1976. Heat and moisture flux in the stratified boundary layer. Q. J. R. Meteorol. Soc., 102: 703–707.

Wyngaard, J.C., 1991. On the maintenance and measurement of scalar fluxes. In: T.J. Schmugge and J.C. André (Editors), Land Surface Evaporation. Springer-Verlag, New York, N.Y., pp. 199–230.

Wyngaard, J.C. and Brost R.A., 1984. Top-down and bottom-up diffusion of a scalar in the convective boundary layer, J. Atmos. Sci. 41: 102–112.

Chapter 5

On the weights of precipitation stations

M. SUGAWARA

Abstract

Even though the Thiessen polygon method for calculating average precipitation has been widely used for a long time, this method is illogical. In principle, the weights of precipitation stations should be determined, not by their geometrical positions, but by the meteorological conditions. For example, in some cases, a station situated near the centre of a basin may be almost useless for discharge forecasting, in spite of its large Thiessen weight. In such cases, due to micro-meteorological conditions, the rainfall in a small area around the station, may have a weak relationship with the areal precipitation over the whole basin.

Some people consider, that by assigning weights to precipitation stations we can find the mean areal precipitation over the basin. However, this is meaningless, because there is no way of measuring the areal precipitation. Instead, there is observed discharge, which is an integral quantity over the basin.

The weights of precipitation stations should be determined from the observed discharge. Even though the observed discharge is an integral quantity over the basin, it seems to be very difficult, even impossible, to derive the areal precipitation over the basin from the observed discharge. Hence, the only way forward is to determine the weights of precipitation stations from a comparison of the observed discharge with the discharge derived from the precipitation at each station. For this purpose, a working runoff model with equal weight applied to each station, is used to calculate the runoff from the precipitation at each station.

The weights of precipitation stations, w_i are found from an optimum estimate of $y(t)$, the observed runoff, using a linear sum $\sum_i w_i x_i(t)$, where $x_i(t)$ is the calculated runoff derived from the ith station. To solve this problem, the method of principal axis transformation and factor analysis, is applied.

1. The Thiessen polygon method is illogical

Even though the Thiessen polygon method has been widely used for a long time, it is illogical and, in reality, cannot give results.

We can see its logical defect by considering two stations situated close to the boundary of a basis, as in Figure 1. Using the Thiessen polygon method, the weight

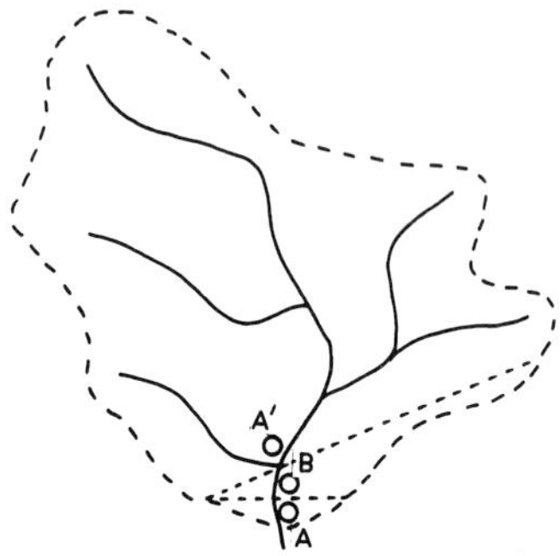

Fig. 1. An example of the case of two adjacent stations, close to the boundary of the basin.

of A is very small and that of B very large. However, if two stations are located close to each other, their weights should be almost equal. If the station A is moved to a A', the weight of A' becomes very large and that of B becomes very small. If the Thiessen polygon method were logical, such an absurdity would not occur.

There are better methods of determining the weights of precipitation stations from their geometrical positions in the basin. For example, precipitation at a point (x,y) can be taken to be a random variable $X(x,y)$ satisfying the following conditions:

(1) $X(x,y)$ is stationary on (x,y).

(2) The correlation coefficient r between $X(x_1,y_1)$ and $X(x_2,y_2)$ is a function of the distance d between the two points:

$$d^2 = (x_1 - x_2)^2 + (y_1 - y_2)^2$$

(3) For example, the correlation coefficient may be given by:

$$r = \exp(-kd)$$

Hence, the areal precipitation over the basin is also a random variable. If there are k precipitation stations at (x_1,y_1), (x_2,y_2), ..., (x_k,y_k), the problem can be formulated as follows:

(1) When w_i $(i = 1, 2, \ldots, k)$ are the weights of precipitation stations, a random variable:

$$\sum_{i=1}^{k} w_i X(x_i,y_i)$$

is an estimate of the areal precipitation over the basin.

(2) The problem is to find the optimum estimate of the areal precipitation over the basin.

Under such an assumption we can determine the weights of precipitation stations, which are different from the Thiessen weights. Moreover, these weights satisfy the desirable condition: when two stations are near, weights of both stations are nearly equal, i.e. weights are continuous functions of the positions of stations.

There is also another way of formulating the problem. Let us assume that the precipitation at a point P, with coordinates and elevation (x,y,h), can be represented by:

$$F(x,y) + kh + X$$

where $F(x,y)$ is a polynomial of 1st, 2nd or higher order and X is a random variable.

In this case, we can also derive optimum weights which are different from the Thiessen weights. This method also satisfies the desirable condition that the weights are continuous functions of the positions of stations.

The author considers, however, that these two methods are merely meaningless, because the weights should not be determined from their geometrical positions, but by considering the meteorological and hydrological conditions.

2. Weights should be determined considering the reliability of observed data

Consideration begins from simple cases. In some cases, unfortunately not so seldom, the observed data may be unreliable. In other cases, data from one station may be similar to data from other stations, or to past data for some period, for example, one month or so, i.e. the missing data were filled in using the similarity with the other data.

Such false or unreliable data are usually identified after several attempts to derive the discharge from precipitation. The results of comparing observed and calculated discharges are sometimes very curious at certain periods in the record. If the given data are found at these times to be similar to other data, we may conclude they are false data.

In some cases, all the data are integer or half integer in value. Such data cannot be reliable.

A precipitation station with false or unreliable data should either be given a small weight or be neglected, i.e. be given a zero weight.

In some other cases, measuring instruments are not good or they may be poorly maintained. In such cases also, the weight of the station must be small.

Even when both the observer and the instrument are satisfactory, in some cases, the surrounding conditions may be unsatisfactory. For example, there may be a large building or tree near the rain gauge. The surrounding conditions may not have been bad initially when the rain gauge was placed, but, in the course of time, they may have deteriorated. In such cases, the weight of the station should be small.

These cases are rather simple and practical, but there is another of a meteorological character. The author has met a case concerning several precipitation stations in and around a basin. There is one station among them situated near the centre of the basin, that is, curiously, nearly useless for flood forecasting, in spite of its apparently good position. A person working for the Dam-Gate Control also knew this fact and has said: "That precipitation station is very curious. Sometimes, it rains heavily at that station and the discharge is small, and sometimes it rains little at that station and the discharge is large. The observer at that station is good, the instruments are good and the surrounding conditions are not bad."

There must be present some local micro-meteorological conditions, such that the local precipitation in a small neighbouring area of the station, is nearly independent of the precipitation over the remaining larger area of the basin.

In contrast to this, there are also cases, where precipitation stations outside the basin are of considerable use for flood forecasting and discharge calculation.

These examples show that the weights of precipitation stations should be determined taking into consideration their usefulness for discharge calculation or flood forecasting. Precipitation stations with false data or unreliable data cannot be useful and so their weights must be small.

3. Weights of precipitation stations should be determined to obtain good results in discharge calculation or flood forecasting

Some people consider that by assigning weights to precipitation stations, we can obtain an optimum estimate of the mean areal precipitation over the basin. However, this approach is rather scholastic and useless, because we cannot observe the mean areal precipitation itself. However, nature integrates the point precipitation over the basin and the integrated value turns into discharge which we can measure. Nature not only makes the areal integral, but also carries out some other operations as well, such as the storage of rain water, runoff from storage, evaporation from storage, time lag, etc.

Even though there are many difficulties, the observed discharge is the only observed data on the areal precipitation. We may pose the problem: how can we derive the mean areal precipitation from the observed discharge? The author considers, that the solution of this problem is nearly impossible.

We may say boldly and simply, that the transformation from precipitation to discharge is an operator of integral character. For example, the tank model is a combined non-linear incomplete integral operator. Therefore, the inverse operator or transformation from discharge to precipitation must have a differential character that is very sensitive to noise. Consequently, the results obtained must be unreliable.

Hence we must select another way forward. Let us assume that precipitation at each station is transformed into calculated runoff by some runoff model and a linear combination of these calculated runoffs is compared with the observed runoff. We seek those weights which yield an optimum estimate of observed runoff as a linear combination of calculated runoff. However, to solve this problem, some runoff model is required, and the weights of precipitation stations are necessary for the runoff model.

Usually, it is better to start from equal weights. A value 1 may be given to stations inside the basin and weights of value 1/2 to stations outside or on the boundary of the basin. Runoff analysis begins using such working weights. In the course of analysis, if some station is found to be unreliable or not representative, it is better to decrease the weight of such a station by subjective judgement.

When the model seems to have converged, i.e. the modeller considers that there is no way to make the model better, the procedure for the choice of weights begins. Therefore, the choice of weights is made in the last stage of runoff analysis. Usually, after the weights have been determined by the mathematical method, described hereafter, runoff analysis begins with these new weights. At this point, the choice of weights is the beginning of the last stage.

4. How to determine the weights of precipitation stations

When the working runoff model has converged, the choice of weights begins. Let the number of precipitation stations be NP. Daily precipitation at the Kth precipitation station is transformed into calculated runoff. Let the desired time series of calculated runoff be:

$$QE(J,K) \qquad (K = 1,\ldots,NP; J = 1,\ldots,LT)$$

where J is the day number and LT is the length of the time series.

Let the time series of observed runoff be:

$$Q(J) \qquad (J = 1,\ldots,LT)$$

Let the weight of the Kth precipitation station be:

$$W(K) \qquad (K = 1,\ldots,NP)$$

then the linear combination:

$$\sum_{K=1}^{NP} W(K) * QE(J,K)$$

is an estimate of $Q(J)$.

The problem is to find an optimum estimate of $Q(J)$. However, such a formulation has some defects. First, the length of the time series is too long. Second, the effect of error of the following type cannot be avoided: Daily precipitation is measured at a fixed observation time, usually 9 o'clock. If heavy rainfall occurs near the observation time, it will be recorded on day J or day $(J + 1)$, by chance. Therefore, a one-day shift of peak between calculated observed discharge is inevitable. A one-day shift of peak causes a large error for two days.

To avoid such defects, monthly or half-monthly averages may be applied. There is the problem that the number of days per month are not the same, but the effect may not be large and may be neglected.

Hereafter, for convenience, we will use normal mathematical notation, instead of FORTRAN, and for simplicity we will consider the case of monthly average data only.

Let the number of precipitation stations be k, and the monthly average runoff calculated from the precipitation at the ith station in the lth month be:

$$x_i(l) \qquad (i = 1, \ldots, k; l = 1, \ldots, n)$$

where n is the length of the time series.

Similarly, let:

$$y(l) \qquad (l = 1, \ldots, n)$$

be the monthly average observed runoff in the lth month.

Time series of calculated and observed monthly average runoff, $\{x_i(l)\}$ and $\{y(l)\}$, respectively, can be represented as vectors:

$$\vec{x}_i = \begin{bmatrix} x_i(1) \\ x_i(2) \\ \vdots \\ x_i(n) \end{bmatrix} \qquad (i = 1, \ldots, k) \qquad \vec{y} = \begin{bmatrix} y(1) \\ y(2) \\ \vdots \\ y(n) \end{bmatrix}$$

The problem is to find the set of optimum weights w_i $(i = 1, \ldots, k)$, which makes the estimate of $\sum_{i=1}^{k} w_i\vec{x}_i$ of $\vec{y}$ optimal.

Originally, it was thought the weights w_i would satisfy the condition:

$$\sum_{i=1}^{k} w_i = 1$$

However, it is neglected hereafter for the following reasons:

(1) Even without the condition $\sum_{i=1}^{k} w_i = 1$, the sum of the weights obtained is usually close to one.

(2) If $\sum_{i=1}^{k} w_i > 1$, using the modified weights $w_i' = w_i/\sum w_i$, will make the estimate $\sum_{i=1}^{k} w_i'\vec{x}_i$, somewhat smaller than $\vec{y}$, on the average.

(3) The adjustment of $\vec{x}_i$ $(i = 1, \ldots, k)$ to make them slightly larger is easy, by making $CE(M)$, the coefficient of evaporation, somewhat smaller or by making $CP(M)$, the coefficient of precipitation, somewhat larger. When $\sum_{i=1}^{k} w_i < 1$, the adjustment is carried out in the opposite sense.

(4) Even if the choice of optimum weights is made subject to the condition $\sum_{i=1}^{k} w_i = 1$, an adjustment of $CE(M)$ or $CP(M)$ is also necessary, for the following reasons:

(a) Equal weights are assumed in the working model.

(b) At some stations, rainfall is larger, on the average and at other stations, smaller.

(c) If the optimum weights obtained show a large value at some station, at which the rainfall is comparatively large, then the estimate $\sum_{i=1}^{k} w_i\vec{x}_i$ will be somewhat larger than $\vec{y}$, on the average.

(d) Therefore, a slight adjustment of $CE(M)$ or $CP(M)$ is also necessary.

5. The problem

The problem is to obtain the optimum estimate of $\vec{y}$ from $\sum_{i=1}^{k} w_i\vec{x}_i$, without imposing the condition $\sum_{i=1}^{k} w_i = 1$.

In the first step, the vectors $\vec{x}_i$ and $\vec{y}$ are normalised to make their norms equal, by either of the following two ways:

(1) The vectors $\vec{x}_i$ and $\vec{y}$ are divided by s_i and s, the root mean square of the components of $\vec{x}_i$ and $\vec{y}$, respectively:

$$\vec{X}_i = \frac{\vec{x}_i}{s_i} \qquad s_i = \left(\sum_{l=1}^{n} \frac{[x_i(l)]^2}{n}\right)^{1/2}$$

$$\vec{Y}_i = \frac{\vec{y}}{s} \qquad s = \left(\sum_{l=1}^{n} \frac{[y(l)]^2}{n}\right)^{1/2}$$

(2) The average is subtracted from every component of each vector and divided by its standard deviation:

$$X_i(l) = \frac{[x_i(l) - m_i]}{s_i} \qquad m_i = \sum_{l=1}^{n} \frac{x_i(l)}{n} \qquad s_i = \left(\sum_{l=1}^{n} \frac{[x_i(l) - m_i]^2}{n}\right)^{1/2}$$

$$Y(l) = \frac{[y(l) - m]}{s} \qquad m = \sum_{l=1}^{n} \frac{y(l)}{n} \qquad s = \left(\sum_{l=1}^{n} [y(l) - m]^2\right)^{1/2}$$

The final results obtained by these two methods are similar, but the former method seems to be slightly better. Therefore, we will use the former method, hereafter.

Then the problem is to find the coefficients c_i which make a norm of the following difference $\vec{d}$ a minimum:

$$\vec{d} = \vec{Y} - \sum_{i=1}^{k} c_i\vec{X}_i$$

If $(\vec{a}, \vec{b})$ is the inner product of two vectors $\vec{a}$ and $\vec{b}$, the vector $\vec{d}$ can be measured by the norm $(\vec{d},\vec{d})$, which is given as:

$$(\vec{d},\vec{d}) = \left(\vec{Y} - \sum_{i=1}^{k} c_i\vec{X}_i,\ \vec{Y} - \sum_{i=1}^{k} c_i\vec{X}_i\right)$$

To determine c_i $(i = 1, \ldots, k)$ which make $(\vec{d},\vec{d})$ a minimum, $(\vec{d},\vec{d})$ is differentiated partially with respect to c_i:

$$\frac{\partial}{\partial c_i}(\vec{d},\vec{d}) = \left(-\vec{X}_i,\ \vec{Y} - \sum_{j=1}^{k} c_j\vec{X}_j\right) + \left(\vec{Y} - \sum_{j=1}^{k} c_j\vec{X}_j,\ -\vec{X}_i\right)$$

$$= 2\left(-(\vec{X}_i,\vec{Y}) + \sum_{j=1}^{k} c_j\,(\vec{X}_i,\vec{X}_j)\right)$$

Putting $\partial/\partial c_i(\vec{d},\vec{d}) = 0$, the following equations are derived:

$$\sum_{j=1}^{k} A_{ij}c_j = B_i \qquad (i = 1, \ldots, k) \tag{1}$$

where:

$$A_{ij} = \frac{(\vec{X}_i,\vec{X}_j)}{n} \qquad B_i = \frac{(\vec{X}_i,\vec{Y})}{n}$$

In the above definition, division by n is not necessary; however, if we include it, A_{ij} can be interpreted geometrically as the cosine of the angle of intersection between $\vec{X}_i$ and $\vec{X}_j$. Likewise, B_i is the cosine of the angle of intersection between $\vec{X}_i$ and $\vec{Y}$.

6. The singular matrix (A_{ij})

Unfortunately, in most cases, the solution of the linear equation (1) is meaningless and useless, showing positive and negative numbers with large absolute values. In most cases, the determinant of the matrix (A_{ij}) is very close to zero, and the roots of equation (1) are given by the ratio of very small numbers, roughly speaking 0/0. Consequently, they are unreliable. The reason why the determinant $|A_{ij}|$ is very close to zero, is the following.

Daily precipitation at stations in or near the basin are usually similar to each other. In other words, they are highly correlated. Therefore, the daily runoffs derived from the precipitation at each station are also similar to each other. Consequently, the time series of calculated mean monthly discharges $x_i(l)$ $(i = 1, \ldots, k)$ are similar to each other, as are the normalized time series $\vec{X}_i$. Hence, the angles of intersection between any two of these vectors are not large. Therefore, all A_{ij} are close to 1, and, consequently, $\det(A_{ij})$ shows a very small value.

Moreover, as the working model has been calibrated completely, the daily runoff derived from the precipitation at each station is similar to observed runoff.

Accordingly, every time series $\vec{X}_i$ $(i = 1, \dots, k)$ is similar to $\vec{Y}$, i.e. the angle of intersection between $\vec{X}_i$ and $\vec{Y}$ is also small, and, consequently, B_i is close to 1.

7. The method of principal axis transformation

To solve a linear system of equations with an almost singular coefficient matrix, we can apply the method of principal axis transformation. First, the previous results are presented in matrix notation. Such a representation is not only convenient for understanding, but also for numerical calculation by computer.

Let A' be the transposed matrix of A. Then $\vec{X}_i'$ and $\vec{Y}'$ are row vectors:

$$\vec{X}_i' = [X_i(1), X_i(2), \dots, X_i(n)]$$

$$\vec{Y}' = [Y(1), Y(2), \dots, Y(n)]$$

Furthermore, A_{ij} and B_i can be represented as:

$$A_{ij} = A_{ji} = \left(\frac{1}{n}\right) \vec{X}_i' \vec{X}_j = \left(\frac{1}{n}\right) \vec{X}_j' \vec{X}_i$$

$$B_i = \left(\frac{1}{n}\right) \vec{X}_i' \vec{Y} = \left(\frac{1}{n}\right) \vec{Y}' \vec{X}_i$$

Let X be a matrix with n rows and k columns, defined as:

$$X = (\vec{X}_1, \vec{X}_2, \dots, \vec{X}_k) = \begin{bmatrix} X_1(1) & X_2(1) & \dots & X_k(1) \\ X_1(2) & X_2(2) & \dots & X_k(2) \\ & \vdots & & \\ X_1(n) & X_2(n) & \dots & X_k(n) \end{bmatrix}$$

The matrix A and the vector $\vec{B}$ can be expressed as:

$$A = \left(\frac{1}{n}\right) X'X = \left(\frac{1}{n}\right) \begin{bmatrix} \vec{X}_1' \\ \vec{X}_2' \\ \vdots \\ \vec{X}_k' \end{bmatrix} (\vec{X}_i, \vec{X}_2, \dots, \vec{X}_k)$$

$$\vec{B} = \left(\frac{1}{n}\right) X'\vec{Y} = \left(\frac{1}{n}\right) \begin{bmatrix} \vec{X}_1' \\ \vec{X}_2' \\ \vdots \\ \vec{X}_k' \end{bmatrix} \vec{Y} \qquad (2)$$

Finally, the vector of weights:

$$\vec{C} = \begin{bmatrix} c_1 \\ c_2 \\ \vdots \\ c_k \end{bmatrix}$$

is introduced, and the linear equation (1) is written as:

$$A\vec{C} = \vec{B} \qquad (1')$$

To find a meaningful solution of this equation, the symmetric matrix A is transformed into diagonal form by the orthogonal transformation T:

$$T'AT = \begin{bmatrix} e_1 & & & \\ & e_2 & & 0 \\ & & \ddots & \\ & 0 & & \\ & & & e_k \end{bmatrix} \qquad (e_1 \geq e_2 \geq \ldots \geq e_k) \qquad (3)$$

$$T'T = E$$

$$T' = T^{-1}$$

where E is the unit matrix.

By multiplying with T on the left-hand side of the above equation (3), we find:

$$AT = T \begin{bmatrix} e_1 & & & \\ & e_2 & & 0 \\ & & \ddots & \\ & 0 & & \\ & & & e_k \end{bmatrix} \qquad (4)$$

Let the ith column of the matrix T be $\vec{t_i}$. The above equation (3) is then equivalent to:

$$A(\vec{t_1}, \vec{t_2}, \ldots, \vec{t_k}) = (\vec{t_1}, \vec{t_2}, \ldots, \vec{t_k}) \begin{bmatrix} e_1 & & & \\ & e_2 & & 0 \\ & & \ddots & \\ & 0 & & \\ & & & e_k \end{bmatrix} = (e_1\vec{t_1}, e_2\vec{t_2}, \ldots, e_k\vec{t_k})$$

namely:

$$A\vec{t_i} = e_i\vec{t_i} \qquad (i = 1, 2, \ldots, k) \qquad (5)$$

This means, $\vec{t_i}$ is the characteristic or eigen vector of A, and e_i is the corresponding characteristic or eigen value.

8. Factor analysis

The method of factor analysis is usually carried out, by neglecting small characteristic values, in such a way, that $e_1, e_2, \ldots, e_{k'}$ are reserved as meaningful and remaining $e_{k'+1}, \ldots, e_k$ are neglected. However, in some cases, the characteristic values decrease gradually, and it is not easy to find the cut-off value for k'.

Therefore, we will consider all k cases, from $k' = 1$ to $k' = k$, and by looking carefully at the results obtained, we will determine k' at the end of the analysis.

Corresponding to the orthogonal transformation T, which transforms A into diagonal form, we will introduce a new vector system $\vec{Z}_i$ $(i = 1, 2, \ldots, k)$, as follows:

$$(\vec{Z}_1, \vec{Z}_2, \ldots, \vec{Z}_k) = (\vec{X}_1, \vec{X}_2, \ldots, \vec{X}_k)T \tag{6}$$

Transposing this result, and using $(AB)' = B'A'$, we find:

$$\begin{bmatrix} \vec{Z}_1' \\ \vec{Z}_2' \\ \vdots \\ \vec{Z}_k' \end{bmatrix} = T' \begin{bmatrix} \vec{X}_1' \\ \vec{X}_2' \\ \vdots \\ \vec{X}_k' \end{bmatrix} \tag{6'}$$

In a similar way to equation (2), we calculate as follows

$$\left(\frac{1}{n}\right) \begin{bmatrix} \vec{Z}_1' \\ \vec{Z}_2' \\ \vdots \\ \vec{Z}_k' \end{bmatrix} (\vec{Z}_1, \vec{Z}_2, \ldots, \vec{Z}_k) = \left(\frac{1}{n}\right) T' \begin{bmatrix} \vec{X}_1' \\ \vec{X}_2' \\ \vdots \\ \vec{X}_k' \end{bmatrix} (\vec{X}_1, \vec{X}_2, \ldots, \vec{X}_k)T$$

and applying the relation (2), this becomes:

$$= T'AT = \begin{bmatrix} e_1 & & & \\ & e_2 & & 0 \\ & & \ddots & \\ & 0 & & \\ & & & e_k \end{bmatrix}$$

We have obtained the result:

$$\left(\frac{1}{n}\right) \begin{bmatrix} \vec{Z}_1' \\ \vec{Z}_2' \\ \vdots \\ \vec{Z}_k' \end{bmatrix} (\vec{Z}_1, \vec{Z}_2, \ldots, \vec{Z}_k) = \begin{bmatrix} e_1 & & & \\ & e_2 & & 0 \\ & & \ddots & \\ & 0 & & \\ & & & e_k \end{bmatrix}$$

Hence, $(1/n)\vec{Z}_i'\vec{Z}_j = \delta_{ij}e_i$, and $\vec{Z}_i$ $(i = 1, \ldots, k)$ form an orthogonal system with $(\vec{Z}_i, \vec{Z}_i)/n = e_i$.

The new vector system $(\vec{Z}_1, \vec{Z}_2, \ldots, \vec{Z}_k)$ is defined by the relation (6), which means:

$$\vec{Z}_i = \sum_j \vec{X}_j T_{ji} = \sum_j T_{ji} \vec{X}_j$$

where T_{ji} is the jth component of the characteristic vector $\vec{t}_i$. Hence, $\vec{Z}_i$ is a weighted average of all $\vec{X}_j$, with the weights given by the components of the vector $\vec{t}_i$.

Usually, the first characteristic vector $\vec{t}_1$ gives nearly equal weights. If all components of the matrix A are a, the first characteristic vector is:

$$\begin{bmatrix} 1/k^{1/2} \\ 1/k^{1/2} \\ \vdots \\ 1/k^{1/2} \end{bmatrix}$$

Frequently, all components of A are nearly equal, and consequently the components of the first characteristic vector are nearly equal. This means, the first factor is close to the simple average with equal weights.

Since (a) the first and second characteristic vectors are orthogonal and (b) the components of the first vector are nearly equal, the components of the second vector must be a mixture of positive and negative values. The distribution of positive and negative components may have a meteorological meaning. For example, positive values may be distributed mostly in the eastern part of a basin and negative values in the western part.

The third characteristic vector is orthogonal to both the first and second vectors, since it is orthogonal to the first vector, its components are a mixture of positive and negative values. And moreover, since it is orthogonal to the second vector, its distribution of positive and negative components is different from that of the second vector. For example, the negative components may be distributed largely in the mountainous parts of the catchment and the positive components largely in the plain. However, in the case of the third, fourth, and higher characteristic vectors, the distribution of positive and negative components on the map, gradually assumes a random pattern, and it becomes difficult to find meteorological meaning. Such characteristic vectors are best neglected.

9. Orthogonal expansion of the runoff

Since the vector system $(\vec{Z}_1, \vec{Z}_2, \ldots, \vec{Z}_k)$ is an orthogonal system, the runoff vector $\vec{Y}$ can be expanded in the form:

$$\vec{Y} = \sum_{i=1}^{k} a_i \vec{Z}_i + \vec{d}$$

where the vector of residuals $\vec{d}$ should be orthogonal to every $\vec{Z}_i$.

The coefficient a_i can be determined by multiplying $\vec{Z}_i'$ on the left side of the above expansion and using the orthogonality property, we find:

$$\left(\frac{1}{n}\right)\vec{Z}_i'\vec{Y} = \left(\frac{1}{n}\right)a_i\vec{Z}_i'\vec{Z}_i = a_ie_i$$

Therefore, a_i is given as:

$$a_i = \left(\frac{1}{n}\right)\frac{\vec{Z}_i'\vec{Y}}{e_i}$$

We can represent this result in matrix form as follows:

$$\vec{a} = \begin{bmatrix} a_1 \\ a_2 \\ \vdots \\ a_k \end{bmatrix} = \left(\frac{1}{n}\right)\begin{bmatrix} \vec{Z}_1'/e_1 \\ \vec{Z}_2'/e_2 \\ \vdots \\ \vec{Z}_k'/e_k \end{bmatrix}\vec{Y} = \left(\frac{1}{n}\right)R\begin{bmatrix} \vec{Z}_1' \\ \vec{Z}_2' \\ \vdots \\ \vec{Z}_k' \end{bmatrix}\vec{Y}$$

In the above equation, R is the diagonal matrix, the diagonal elements of which are reciprocals of the characteristic values, namely $1/e_1$, $1/e_2, \ldots, 1/e_k$. Applying the relation (6′) to the above equation, we obtain:

$$\vec{a} = \begin{bmatrix} a_1 \\ a_2 \\ \vdots \\ a_k \end{bmatrix} = \left(\frac{1}{n}\right)RT'\begin{bmatrix} \vec{X}_1' \\ \vec{X}_2' \\ \vdots \\ \vec{X}_k' \end{bmatrix}\vec{Y} = RT'B$$

This means:

$$a_i = \sum_{l=1}^{k}\sum_{m=1}^{k}\left(\frac{\delta_{il}}{e_l}\right)T_{lm}'B_m = \sum_{l=1}^{k}\sum_{m=1}^{k}\left(\frac{\delta_{il}}{e_l}\right)T_{ml}B_m = \sum_{m=1}^{k}\left(\frac{1}{e_i}\right)T_{mi}B_m$$

$$= \sum_{m=1}^{k}\frac{T_{mi}B_m}{e_i}$$

Neglecting the characteristic values smaller than $e_{k'}$, $\vec{Y}$ is approximated by $\sum_{i=1}^{k'} a_i\vec{Z}_i$, which can be represented as:

$$(\vec{Z}_1,\vec{Z}_2,\ldots\vec{Z}_k)\begin{bmatrix} a_1 \\ a_2 \\ \vdots \\ a_{k'} \\ 0 \\ \vdots \\ 0 \end{bmatrix} = (\vec{Z}_1,\vec{Z}_2,\ldots,\vec{Z}_k)R(k')T'\vec{B}$$

In the above equation, $R(k')$ is the diagonal matrix, the diagonal elements of which are $1/e_1, 1/e_2, \ldots, 1/e_{k'}, 0, \ldots, 0$.

10. The precipitation weights

To obtain the weights of each precipitation station, $\sum_{i=1}^{k'} a_i \vec{Z}_i$ must be transformed into the linear sum of $\vec{X}_i$. For this purpose, we apply the relation (6) to the above equation:

$$\sum_{i=1}^{k'} a_i \vec{Z}_i = (\vec{X}_1, \vec{X}_2, \ldots, \vec{X}_k) T R(k') T' \vec{B}$$

The right-hand side of this equation is the optimum estimate of $\vec{Y}$, represented as $\sum_{i=1}^{k} c_i(k') \vec{X}_i$, where $c_i(k')$ ($i = 1, 2, \ldots, k$) are the weights of the normalized variables $\vec{X}_i$ approximating $\vec{Y}$, in the case of k'. From the above expression, the weights are given as:

$$\vec{C}(k') = \begin{bmatrix} c_1(k') \\ c_2(k') \\ \vdots \\ c_k(k') \end{bmatrix} = T R(k') T' \vec{B}$$

Though this expression can be used for numerical calculation by computer, we will also give the usual expression without using matrix notation.

Let S be $T R(k')$, then:

$$S_{ij} = \begin{cases} T_{ij}/e_j & (j \leq k') \\ 0 & (j > k') \end{cases}$$

With the use of S, $\vec{C}(k')$ is given by:

$$\vec{C}(k') = S T' \vec{B} = \left(\sum_{l=1}^{k} \sum_{m=1}^{k} S_{il} T'_{lm} B_m \right) = \left(\sum_{l=1}^{k} \sum_{m=1}^{k} S_{il} T_{ml} B_m \right)$$

$$= \left(\sum_{l=1}^{k'} \sum_{m=1}^{k} \left(\frac{T_{il}}{e_l} \right) T_{ml} B_m \right)$$

Namely:

$$C_i(k') = \sum_{l=1}^{k'} \sum_{m=1}^{k} \frac{T_{il} T_{ml} B_m}{e_l}$$

Finally, the mean square error of the k'th optimum estimate of the runoff $\sum_{i=1}^{k} c_i(k')\vec{X}_i = \sum_{i=1}^{k'} a_i\vec{Z}_i$ must be calculated as the norm $[\vec{d}(k'), \vec{d}(k')]$, where:

$$\vec{d}(k') = \vec{Y} - \sum_{i=1}^{k'} a_i\vec{Z}_i$$

the vector of residual errors. Since $\vec{d}(k')$ is orthogonal to $\sum_{i=1}^{k'} a_i\vec{Z}_i$, $[\vec{d}(k'), \vec{d}(k')]$ can be calculated using the Pythagorean formula as follows:

$$\left(\frac{1}{n}\right)[\vec{d}(k'), \vec{d}(k')] = \left(\frac{1}{n}\right)(\vec{Y}, \vec{Y}) - \left(\frac{1}{n}\right)\left(\sum_{i=1}^{k'} a_i\vec{Z}_i, \sum_{i=1}^{k'} a_i\vec{Z}_i\right)$$

$$= \left(\frac{1}{n}\right)(\vec{Y}, \vec{Y}) - \left(\frac{1}{n}\right)\sum_{i,j=1}^{k'} a_i a_j(\vec{Z}_i, \vec{Z}_j) = \left(\frac{1}{n}\right)(\vec{Y}, \vec{Y}) - \sum_{i,j=1}^{k'} a_i a_j \delta_{ij} e_i$$

$$= \left(\frac{1}{n}\right)(\vec{Y}, \vec{Y}) - \sum_{i=1}^{k'} a_i^2 e_i$$

where:

$$a_i = \sum_{m=1}^{k} T_{mi} B_m e_i$$

The weights $\vec{C}(k')$ $(k' = 1, \ldots, k)$ have been found for the normalized variables $\vec{X}_i$ and $\vec{Y}$. They will now be transformed into the weights $\vec{W}(k')$ for $\vec{x}_i$ and $\vec{y}$.

Since:

$$\vec{y} = \sum_{i=1}^{k} c_i(k')\vec{X}_i + \vec{d}(k')$$

means:

$$\frac{\vec{y}}{s} = \sum_{i=1}^{k} c_i(k')\left(\frac{\vec{x}_i}{s_i}\right) + \vec{d}(k')$$

therefore:

$$\vec{y} = \sum_{i=1}^{k} c_i(k')\left(\frac{s}{s_i}\right)\vec{x}_i + s\vec{d}(k')$$

Hence, the weights $w_i(k')$ are given by:

$$w_i(k') = c_i(k')\left(\frac{s}{s_i}\right)$$

Usually, $\sum_{i=1}^{k} w_i(k')$ is close to 1.

11. The determination of the cut-off point k'

The last problem is the choice of k' by subjective judgement, using the following rules:

(1) Looking at the characteristic values, if $e_{k'+1}$ is much smaller than $e_{k'}$, k' is determined.

(2) Looking at the components of the ith characteristic vector $\vec{t}(i)$, if the distribution of positive and negative components on the map of the basin seems to have meteorological meaning, the vector $\vec{t}(i)$ must be retained. If the distribution seems to be random, the characteristic vector must be neglected.

(3) Looking at the weights $w_i(k')$ ($i = 1, \ldots, k$), if a negative weight is found, this k' should be neglected.

(4) Looking at the mean square error of k'th estimate, if the k'th one shows a large decrease in error compared with the previous one, and thereafter the decrease in error is rather small, this k' must be selected.

Usually, these conditions are not in contradiction with each other, and so, the judgement is rather objective and not difficult.

References

Sugawara, M., 1979. Automatic calibration of the tank model. Hydrol. Sci. Bull. 24(3): 375–388.

Chapter 6

A Neyman-Scott shot noise model for the generation of daily streamflow time series

P. S.P. COWPERTWAIT and P.E. O'CONNELL

Abstract

This paper introduces a Neyman-Scott shot noise model with a plausible physical structure for the generation of synthetic streamflow data. The model has two components: (a) an event process which is represented by a Neyman-Scott instantaneous pulses model, and (b) a linear response process which consists of two linear reservoirs in parallel. A proportion of each pulse is routed through a "fast" linear reservoir to represent surface runoff, and the remainder through a "slow" linear reservoir to represent baseflow; the latter provides a means of preserving persistence at the monthly level. The second-order moments for the model are derived and used to fit the model to two 25-year data sets taken from different catchments in the UK: the upper River Thames at Eynsham which has a substantial groundwater flow component, and the River Coquet at Morwick which is a more responsive surface runoff catchment. The model is validated through a comparison of simulated and historical statistics not used in fitting, such as the distribution of annual maximum daily flows in each year and the proportions of flow below certain levels. The simulated proportions match the historical proportions closely for both sites, although the model shows a tendency to over-estimate the extreme values, particularly for the Thames. However, overall the model fits both sites reasonably well, and, in particular, plots of the simulated time series exhibit the main characteristics present in the historical plots. Some possible further developments of the model are outlined.

1. Introduction

A tribute of this kind would not be complete without at least one contribution in celebration of the mathematical elegance of linear system theory in hydrology, so powerfully demonstrated many years ago by Dooge (1959). Today, linear systems theory still occupies a dominant position in both theoretical and applied hydrology, and will doubtless continue to do so for many years to come. Here, the theory of clustered Poisson processes is combined with some linear system theory to yield a physically plausible stochastic model of daily streamflow.

One of the main purposes of stochastic streamflow modelling is to generate synthetic flows for use in water resource system simulation studies. However, streamflow exhibits distinctly different properties when averaged over different time scales. For monthly and annual flows, classical ARMA modelling approaches (Box and Jenkins, 1970) can be justified, and the literature abounds with applications of such models (e.g. O'Connell, 1974; Lawrence and Kottegoda, 1977; McLeod and Hipel, 1978; Salas and Obsekera, 1982; Lungu and Sefe, 1991). However, on shorter timescales, streamflow exhibits the influences of individual rainstorms which generate the characteristic rapid rises and slow recessions associated with the response of streamflow to rainfall. The problem of modelling such features stochastically is more difficult than modelling monthly and annual averages, and has received rather less attention in the literature. Weiss (1977) proposed that shot noise models could be used to reproduce these features.

In its simplest form, a shot noise model is characterized by a series of instantaneous pulses which occur along the time scale according to a simple Poisson process; in loose terms these pulses might be taken to correspond to effective rainfall values. Each pulse is routed through a linear reservoir to generate a response, and the responses aggregated to yield the continuous time shot noise process. The process is then averaged to yield a discrete time process, referred to by Weiss (1977) as a Single Shot Noise (SSN) Model. Weiss found that a SSN model of daily streamflow was unsatisfactory, primarily because of its inability to model baseflow, and proposed a Double Shot Noise (DSN) Model consisting of the sum of two independent shot noise processes with "fast" and "slow" responses. However, the assumption of independence of the two processes leads to physically unrealistic streamflow behaviour, since pulses in the slow process can occur without any corresponding pulses in the fast process. Furthermore, the assumption of a temporally independent Poisson process has been criticized by Vandewiele and Dom (1989) who proposed an alternative model based on a Markov process. The limitations of an independent Poisson process in modelling rainfall time series have also been recognised, and a clustered point process (such as the Neyman-Scott model) has been used successfully to model the dependence inherent in rainfall events (e.g. Rodriguez-Iturbe et al., 1987; Cowpertwait, 1991). It therefore seems natural to generalize the Poisson shot noise model to allow for the possibility of the event process being dependent, while at the same time attempting to achieve a more physically realistic response process.

2. Definition of a second order Neyman-Scott shot noise (NSSN) model

In defining the model, generality is aimed for, so that the model equations for more simple point process streamflow models can be obtained from the equations for the more general version of the model.

Consider a single geographical location on a river. It is assumed that storm origins arrive at the location according to a Poisson process of rate λ (per hour),

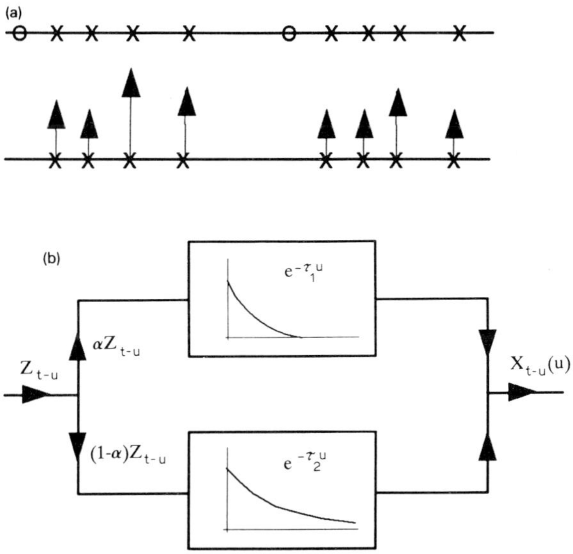

Fig. 1. (a) Schematic of the Neyman-Scott instantaneous pulses model showing storm origins (o) and cell origins (×). (b) Schematic of the proportional-input linear system model used in the NSSN model.

and that each storm origin gives rise to a random number C (where $C \geq 1$) of cell origins. The cell origins are independently displaced from the storm origin by times which are exponentially distributed with parameter β (per hour), no cell origin being located at the storm origin. A random instantaneous pulse Z in the streamflow with mean μ_Z is associated independently with each cell origin. This model of the pulse process (Figure 1a) is referred to as the Neyman-Scott instantaneous pulses model (Rodriguez-Iturbe et al., 1984). Since the model is to be used to model daily streamflow, the assumption of instantaneous pulses is reasonable, since the effective rainfalls associated with individual storms will normally not exceed 1 day in duration, although they may, of course, be split between two successive days.

The response to each pulse Z is modelled as follows. A proportion αZ is routed through a linear reservoir with response function $e^{-\tau_1 u}$ and the remainder $(1-\alpha)Z$ is routed through a second linear reservoir with response function $e^{-\tau_2 u}$ where $\tau_1 \gg \tau_2$ (Figure 1b). This is a linear system which satisfies the properties of proportionality and superposition; the "fast" response with time constant τ_1 can be interpreted as surface runoff, while the slow response can be interpreted as baseflow. Weiss (1977) considered this response model in conjunction with an independent Poisson model of the event process. Since the model can be described by a second order differential equation, this gave rise to the title "Second Order Shot Noise (SOSN) model"; however, Weiss did not explore its properties fully or apply it to daily streamflow.

A second order Neyman-Scott Shot Noise model is now defined mathematically as follows. If $X_{t-u}(u)$ denotes the streamflow response at time t due to a cell with origin at time $t-u$, then:

$$X_{t-u}(u) = \alpha Z_{t-u} e^{-\tau_1 u} + (1-\alpha) Z_{t-u} e^{-\tau_2 u}$$

where Z_{t-u} is the instantaneous pulse occurring at the cell origin.

Furthermore, it is assumed that the total flow $Y(t)$ at time t is the sum of all the surface flow and baseflow contributions due to all cells with origins preceding t, i.e.:

$$Y(t) = \int_{u=0}^{\infty} X_{t-u}(u)\, \mathrm{d}N(t-u)$$

where:

$$\mathrm{d}N(t-u) = 1 \text{ if there is a cell origin at time } t-u$$
$$= 0 \text{ otherwise}$$

i.e.:

$$Y(t) = \int_{u=0}^{\infty} \{\alpha Z_{t-u} e^{-\tau_1 u} + (1-\alpha) Z_{t-u} e^{-\tau_2 u}\}\, \mathrm{d}N(t-u) \tag{1}$$

For brevity, the process $Y(t)$ will be referred to subsequently as the Neyman-Scott Shot Noise model.

3. Some moments of the Neyman-Scott shot noise model

The mean of the process $Y(t)$ follows from equation (1) as:

$$E[Y(t)] = \int_{u=0}^{\infty} E\{\alpha Z_{t-u} e^{-\tau_1 u} + (1-\alpha) Z_{t-u} e^{-\tau_2 u}\}\, \lambda \mu_c\, \mathrm{d}u$$

where μ_c denotes the mean number of cells per storm, i.e.:

$$E[Y(t)] = \lambda \mu_c \mu_z \left(\frac{\alpha}{\tau_1} + \frac{(1-\alpha)}{\tau_2} \right) \tag{2}$$

where μ_z is the mean of the pulses.

The autocovariance of $Y(t)$ at lag k is given by:

$$C_y(k) = \mathrm{Cov}\{Y(t), Y(t+k)\}$$

$$= \int_0^{\infty}\int_0^{\infty} E\{X_{t-u}(u) X_{t+k-v}(v)\}\, \mathrm{Cov}\{\mathrm{d}N(t-u), \mathrm{d}N(t+k-v)\}$$

$$= \int_k^\infty E\left\{[\alpha e^{-\tau_1 v} + (1-\alpha)e^{-\tau_2 v}] [\alpha e^{-\tau_1(v-k)} + (1-\alpha)e^{-\tau_2(v-k)}] Z^2_{t+k-v}\right\} \mathrm{Var}\{\mathrm{d}N(t+k-v)\}$$

$$+ \int\int_{u \neq v-k} E\left\{[\alpha e^{-\tau_1 u} + (1-\alpha)e^{-\tau_2 u}] [\alpha e^{-\tau_1 v} + (1-\alpha)e^{-\tau_2 v}]\right\} \mathrm{Cov}\{\mathrm{d}N(t-u), \mathrm{d}N(t+k-v)\} \quad (3)$$

From Rodriguez-Iturbe et al. (1987):

$$\mathrm{Cov}\{\mathrm{d}N(t-u), \mathrm{d}N(t+k-v)\}$$

$$= \lambda\mu_c\left[\delta(u+k-v) + \frac{1}{2}\mu_c^{-1}E\{C^2 - C\}\beta e^{-\beta(u+k-v)}\right]\mathrm{d}u\,\mathrm{d}v \quad \text{if } v < u+k$$

$$= \lambda\mu_c\left[\delta(v-u-k) + \frac{1}{2}\mu_c^{-1}E\{C^2 - C\}\beta e^{-\beta(v-u-k)}\right]\mathrm{d}u\,\mathrm{d}v \quad \text{if } v > u+k$$

where δ denotes the Dirac delta function. Therefore:

$$C_y(k) = \int_k^\infty E\left\{[\alpha e^{-\tau_1 v} + (1-\alpha)e^{-\tau_2 v}] [\alpha e^{-\tau_1(v-k)} + (1-\alpha)e^{-\tau_2(v-k)}] Z^2_{t+k-v}\right\} \lambda\mu_c\,\mathrm{d}v$$

$$+ \int_0^\infty\int_0^{u+k} E\left\{[\alpha e^{-\tau_1 u} + (1-\alpha)e^{-\tau_2 u}] [\alpha e^{-\tau_1 v} + (1-\alpha)e^{-\tau_2 v}]\right\} \frac{\lambda}{2}E\{C^2 - C\}\beta e^{-\beta(u+k-v)}\,\mathrm{d}v\,\mathrm{d}u$$

$$+ \int_0^\infty\int_{u+k}^\infty E\left\{[\alpha e^{-\tau_1 u} + (1-\alpha)e^{-\tau_2 u}] [\alpha e^{-\tau_1 v} + (1-\alpha)e^{-\tau_2 v}]\right\} \frac{\lambda}{2}E\{C^2 - C\}\beta e^{-\beta(v-u-k)}\,\mathrm{d}v\,\mathrm{d}u \quad (4)$$

Hence, after integrating the various terms in (4), the following expression is obtained:

$$C_y(k) = A_1 e^{-\tau_1 k} + A_2 e^{-\tau_2 k} + B[C_1 e^{-\tau_1 k} + C_2 e^{-\tau_2 k} + De^{-\beta k}] \quad (5)$$

where:

$$A_1 = \frac{\lambda\mu_c E[Z^2]}{2\tau_1\tau_2(\tau_1+\tau_2)}\alpha\tau_2[\alpha\tau_2 + (2-\alpha)\tau_1]$$

$$A_2 = \frac{\lambda\mu_c E[Z^2]}{2\tau_1\tau_2(\tau_1+\tau_2)}(1-\alpha)\tau_1[(1-\alpha)\tau_1 + (1+\alpha)\tau_2]$$

$$B = \frac{\lambda\beta\mu_Z^2 E[C^2 - C]}{2\tau_1\tau_2(\tau_1 + \tau_2)(\beta^2 - \tau_1^2)(\beta^2 - \tau_2^2)}$$

$$C_1 = \alpha\beta\tau_2(\beta^2 - \tau_2^2)[\alpha(\tau_2 - \tau_1) + 2\tau_1]$$

$$C_2 = (1 - \alpha)\beta\tau_1(\beta^2 - \tau_1^2)[\alpha(\tau_2 - \tau_1) + \tau_1 + \tau_2]$$

$$D = \tau_1\tau_2(\tau_1 + \tau_2)[\alpha(\tau_2 - \tau_1) + \tau_1 - \beta][\alpha(\tau_2 - \tau_1) + \tau_1 + \beta]$$

Streamflow records tend to be available as average flows over some time increment (e.g. hours or days). Let $Y_i^{(h)}$ denote the average flow for the ith interval of length h (in hours); then:

$$Y_i^{(h)} = \frac{1}{h}\int_{(i-1)h}^{ih} Y(t)\,\mathrm{d}t \tag{6}$$

The second order moments of this aggregated process can be obtained from:

$$\left.\begin{aligned} E\left[Y_i^{(h)}\right] &= E[Y(t)] \\ \mathrm{Var}\left[Y_i^{(h)}\right] &= \frac{2}{h^2}\int_0^h (h-u)C_y(u)\,\mathrm{d}u \\ \mathrm{Cov}\left[Y_i^{(h)}, Y_{i+k}^{(h)}\right] &= \frac{1}{h^2}\int_{-h}^{h}(h - |v|)C_y(kh + v)\,\mathrm{d}v \end{aligned}\right\} \tag{7}$$

(see Rodriguez-Iturbe et al., 1987, equation 2.14).

Using equations (7) and (5) the following results are obtained:

$$\begin{aligned} \mathrm{Var}\,[Y_i^{(h)}] = {} & \frac{2}{h^2\tau_1^2}[\tau_1 h - 1 + e^{-\tau_1 h}][A_1 + BC_1] \\ & + \frac{2}{h^2\tau_2^2}[\tau_2 h - 1 + e^{-\tau_2 h}][A_2 + BC_2] + \frac{2BD}{h^2\beta^2}[\beta h - 1 + e^{-\beta h}] \end{aligned} \tag{8}$$

$$\begin{aligned} \mathrm{Cov}\left[Y_i^{(h)}, Y_{i+k}^{(h)}\right] = {} & \frac{1}{h^2\tau_1^2}e^{-\tau_1 h(k-1)}(1 - e^{-\tau_1 h})^2[A_1 + BC_1] \\ & + \frac{1}{h^2\tau_2^2}e^{-\tau_2 h(k-1)}(1 - e^{-\tau_2 h})^2[A_2 + BC_2] + \frac{BD}{h^2\beta^2}e^{-\beta h(k-1)}(1 - e^{-\beta h})^2 \end{aligned} \tag{9}$$

In equations (2), (8) and (9) the distributions for the cell intensity Z and the number of cells per storm C have not been specified. An obvious choice for these, and this is the choice made in fitting the NSSN model below, is to take Z as an expo-

TABLE 1

Some special cases of the Neyman-Scott Shot Noise Model

Combination	Name of model
$C = \alpha = 1$	Single Shot Noise (SSN) model
$C = 1, \quad 0 < \alpha < 1$	Second Order Shot Noise (SOSN) model
$C > 1, \quad \alpha = 1$	Single NSSN model
$C > 1, \quad 0 < \alpha < 1$	Second Order NSSN model

nential random variable, and $(C-1)$ as a Poisson random variable [a mean of $(C-1)$, rather than C, ensures that at least one raincell occurs with every storm origin].

4. Some special cases

The following combinations of C and α in equations (2), (8) and (9) give rise to some special cases of the NSSN model as summarized in Table 1. The SSN and SOSN models have been studied by Weiss (1977) and O'Connell and Jones (1979). In addition, Double Shot Noise (DSN) models, where pulses in the surface and baseflow processes occur as independent Poisson processes, have also been considered (Weiss, 1977). As already noted, the DSN model is not appealing from a physical standpoint, and so the mathematical development of a Neyman-Scott Double Shot Noise model has not been carried out in this paper, although clearly the mathematics would be more straightforward than for the NSSN model because of the independence of the surface runoff and baseflow processes.

5. Fitting the NSSN model to historical daily flows

In order to test the performance of the NSSN model, two catchments with contrasting hydrological regimes were selected. The River Coquet at Morwick in North East England (area 570 km^2) is predominantly a surface runoff catchment while the response of the River Thames at Eynsham in the South of England (area 1616 km^2) is strongly influenced by the Chalk aquifer. Daily flows for the periods 1952–1976 and 1965–1989 were available for the Rivers Thames and Coquet, respectively (25 years in both cases). Data for two typical years are plotted in Figure 2 for the River Thames and in Figure 3 for the River Coquet; for the Thames, there is evidence of a larger proportion of groundwater runoff and more slowly decaying recessions. This leads to more persistence from one month to the next, suggesting that high-lag daily flow autocorrelations should be used in fitting the NSSN model to the Thames data.

A traditional approach to fitting models of the NSSN type is to estimate m low-order moments from the available data, where m is the number of model parameters, and to solve the moment equations for the parameter values. The equations are

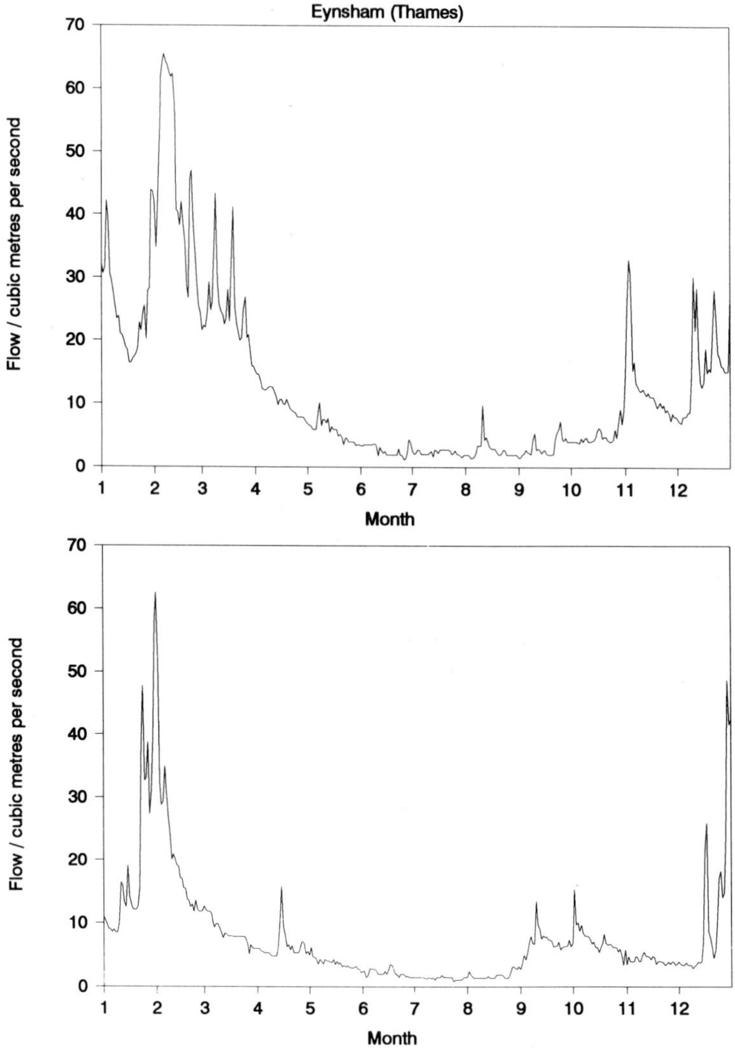

Fig. 2. Plots of two years of historical daily flows: River Thames at Eynsham.

usually non-linear in the parameters and, while an exact solution may be found, the resulting estimates may not always be physically reasonable. A more flexible approach has been adopted here in which a sum of squares function, defined as:

$$SS = \sum_{j=1}^{n} W_j \left(1 - \frac{M_j}{\hat{M}_j}\right)^2 \tag{10}$$

is minimized, where M_j is the model moment, $\hat{M}_j$ is its sample estimate, and W_j allows greater weight to be given to fitting some moments relative to others. The

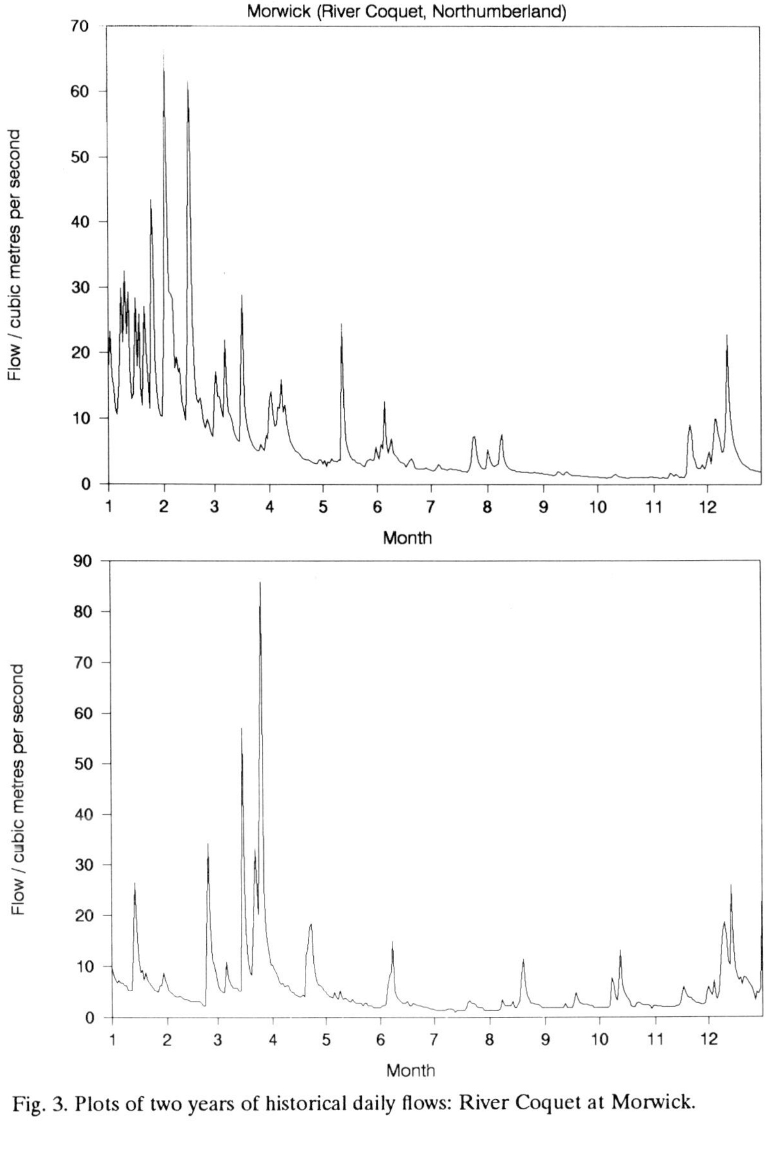

Fig. 3. Plots of two years of historical daily flows: River Coquet at Morwick.

function SS is minimized by conducting a search in the model parameter space, where n, the number of moments used in fitting, is equal to or greater than m, the number of parameters. Bounds can be placed on each of the parameters so that a priori knowledge of physically realistic limits for the parameter estimates can be imposed, while matching as far as possible the properties used in fitting.

For each calendar month, the following statistics were extracted from each historical data set: the daily mean ($\bar{x}$), the daily variance (s^2), and the daily autocorrelations at lags 1, 2, 3, 4, 5, 6, 12 and 24 days [$\hat{\rho}(1)$, $\hat{\rho}(2)$, $\hat{\rho}(3)$, $\hat{\rho}(4)$,

$\hat{\rho}(5)$, $\hat{\rho}(6)$, $\hat{\rho}(12)$ and $\hat{\rho}(24)$, respectively]. The variances and autocovariances were found using the overall sample mean ($\bar{x}$) in the calculations to avoid bias (e.g. see Trenberth, 1984).

For the River Coquet, the parameters of the NSSN model were estimated for each calendar month by minimizing the sum of squares function (10) with:

$$M_1 = \mu = E[Y(t)]$$

$$M_2 = \sigma^2 = \mathrm{Var}\,[Y_i^{(24)}]$$

$$M_{k+2} = \rho(k) \qquad \mathrm{k} = 1, 2, \ldots, 6$$

where:

$$\rho(k) = \frac{\mathrm{Cov}\,[Y_i^{(24)}, Y_{i+k}^{(24)}]}{\mathrm{var}\,[Y_i^{(24)}]}$$

and expressions for the mean, variance and covariances are given by equations (2), (8) and (9). For the Thames data set, higher lag covariances were used in the fitting procedure because of the greater persistence evident in the time series plots:

$$M_1 = \mu$$

$$M_2 = \sigma^2$$

$$M_{k+2} = \rho(s), \qquad (k, s) = (1,1), (2,3), (3,6), (4,12), (5,24)$$

For both sites, the weights W_j were all set equal to unity, thus giving equal weight to all terms included in the sum of squares function. This results from the use of a ratio in each term.

The minimisations were carried out using a NAG (1977) routine called E04JAF, which enabled bounds to be placed on the parameters. The bounds were chosen to ensure that the baseflow had a slow decay rate, the surface flow a rapid decay rate, and that at least some of the flow appeared as baseflow and surface flow. The parameters were constrained as follows:

$$10^{-3} \leq \lambda \leq 10^{-1}; \quad 10^{-3} \leq \beta \leq 10; \quad 1 \leq \mu_c \leq 20; \quad 10^{-2} \leq \mu_z^{-1} \leq 10;$$

$$10^{-2} \leq \tau_1 \leq 10; \quad 10^{-8} \leq \tau_2 \leq 10^{-3}; \quad 0.01 \leq \alpha \leq 0.99$$

When the parameters were estimated it was found that β, the mean waiting time for the cell origins, usually reached its upper bound of 10 (per hour). This implies that the data sets do not contain sufficient information to enable the "within storm" cellular structure to be found. This is attributable to the level of aggregation of the data sets being too large, since raincells tend to be separated by periods of less than one day, and to the smoothing carried out by the catchment

TABLE 2

NSSN parameter estimates for the River Thames at Eynsham

Month	λ (hr^{-1})	β (hr^{-1})	μ_c	μ_z (m^3 s^{-1})	τ_1 (hr^{-1})	τ_2 (hr^{-1})	α
Jan	0.00275	10.0	13.8	2.52	0.0100	0.00100	0.795
Feb	0.00187	10.0	10.0	3.91	0.0121	0.000632	0.815
Mar	0.00613	10.0	2.11	7.97	0.0100	0.00100	0.890
Apr	0.00803	10.0	1.00	7.14	0.0100	0.00100	0.850
May	0.00385	10.0	4.95	2.03	0.0105	0.00100	0.823
Jun	0.00136	10.0	7.01	3.41	0.0186	0.00100	0.803
Jul	0.00120	10.0	14.2	1.44	0.0100	0.00100	0.884
Aug	0.00201	10.0	8.35	1.21	0.0610	0.00100	0.828
Sep	0.00100	10.0	14.3	1.70	0.0440	0.000749	0.864
Oct	0.00100	10.0	7.33	3.62	0.0106	0.00100	0.683
Nov	0.00100	10.0	1.48	26.2	0.0130	0.000706	0.753
Dec	0.00113	10.0	18.6	2.54	0.0100	0.00100	0.649

TABLE 3

NSSN parameter estimates for the River Coquet at Morwick

Month	λ (hr^{-1})	β (hr^{-1})	μ_c	μ_z (m^3 s^{-1})	τ_1 (hr^{-1})	τ_2 (hr^{-1})	α
Jan	0.00487	10.0	1.00	44.25	0.0356	0.00100	0.958
Feb	0.00676	10.0	1.00	24.15	0.0156	0.000535	0.990
Mar	0.00514	10.0	1.00	33.90	0.0270	0.00100	0.964
Apr	0.00209	10.0	1.00	30.67	0.0297	0.00100	0.898
May	0.00169	10.0	1.00	35.46	0.0637	0.00100	0.925
Jun	0.00332	10.0	1.00	11.49	0.0401	0.00100	0.930
Jul	0.00100	10.0	1.00	34.13	0.0456	0.000354	0.974
Aug	0.00165	10.0	1.00	54.64	0.0506	0.00100	0.970
Sep	0.00155	10.0	1.00	43.86	0.0563	0.00100	0.953
Oct	0.00100	10.0	1.00	78.13	0.0492	0.000742	0.944
Nov	0.00157	10.0	1.00	69.93	0.0382	0.00100	0.924
Dec	0.0110	10.0	1.00	18.45	0.0186	0.00100	0.990

on individual raincells. Hence, β was fixed at its upper bound of 10 (per hour), which implies that all the cells occur close to the storm origin. The parameter estimates for the River Thames and the River Coquet are given in Tables 2 and 3, respectively.

For the River Coquet, the mean number of cells per storm was found to be 1 for most months which implies that the more simple SOSN model can describe this data set adequately (a mean number of cells per storm of 1 implies that the number of cells for each storm is constant and equal to 1, because $(C - 1)$ is assumed to follow a Poisson distribution). Hence, for this data set, both β and μ_c were fixed.

TABLE 4

Comparison of historical and fitted statistics for the River Thames at Eynsham

Month		$\bar{x}$	s^2	$\hat{\rho}(1)$	$\hat{\rho}(3)$	$\hat{\rho}(6)$	$\hat{\rho}(12)$	$\hat{\rho}(24)$
Jan	hist	27.24	297.94	0.928	0.764	0.600	0.355	0.256
	fitted	27.24	297.94	0.914	0.715	0.537	0.376	0.259
Feb	hist	26.42	254.31	0.914	0.730	0.564	0.456	0.364
	fitted	26.42	254.31	0.912	0.722	0.571	0.453	0.364
Mar	hist	20.53	178.80	0.910	0.760	0.646	0.433	0.115
	fitted	20.53	178.79	0.887	0.630	0.410	0.233	0.143
Apr	hist	13.46	54.63	0.840	0.689	0.521	0.372	0.168
	fitted	13.46	54.63	0.899	0.668	0.466	0.296	0.194
May	hist	9.91	37.10	0.885	0.649	0.466	0.449	0.210
	fitted	9.91	37.10	0.903	0.684	0.496	0.337	0.231
Jun	hist	7.79	48.38	0.872	0.618	0.583	0.419	0.301
	fitted	7.79	48.38	0.880	0.656	0.519	0.420	0.313
Jul	hist	4.98	34.33	0.967	0.819	0.541	0.389	0.123
	fitted	4.98	34.33	0.889	0.636	0.418	0.243	0.151
Aug	hist	3.77	5.54	0.857	0.716	0.671	0.616	0.415
	fitted	3.77	5.54	0.854	0.724	0.669	0.579	0.434
Sep	hist	4.40	16.67	0.859	0.601	0.519	0.650	0.485
	fitted	4.91	15.46	0.827	0.642	0.589	0.527	0.425
Oct	hist	8.49	109.02	0.939	0.819	0.661	0.452	0.486
	fitted	10.12	93.43	0.936	0.787	0.648	0.507	0.367
Nov	hist	14.75	266.40	0.939	0.795	0.641	0.464	0.554
	fitted	15.86	253.49	0.927	0.768	0.642	0.533	0.427
Dec	hist	22.19	334.21	0.929	0.793	0.709	0.522	0.385
	fitted	22.19	334.21	0.943	0.807	0.675	0.533	0.386

Some notable features of the parameter estimates are as follows. Firstly, the bounds for the decay rate parameters τ_1 and τ_2 are often reached. At first this suggested that a satisfactory minimum had not been found. However, the historical statistics used in the fitting procedure were found to be closely matched by the model and so this did not cause concern (see Tables 4 and 5). Secondly, on comparing Tables 2 and 3 it is evident that the proportion of flow α appearing as surface runoff is greater for the River Coquet than for the Thames. This is a physically realistic result as the Coquet catchment is known to be more responsive.

6. Model validation

To validate the model, 25 years of daily streamflow data were simulated for both sites using the parameter estimates in Tables 2 and 3. The proportion of time (over the whole 25 years)that the flow was less than specified levels, in the range 0.5 to 30 $m^3\ s^{-1}$, was found for both the historical and simulated records. These proportions were plotted against flow level for the River Thames (Figure 4) and the

TABLE 5

Comparison of historical and fitted statistics for the River Coquet at Morwick

Month		$\bar{x}$	s^2	$\hat{\rho}(1)$	$\hat{\rho}(2)$	$\hat{\rho}(3)$	$\hat{\rho}(4)$	$\hat{\rho}(5)$	$\hat{\rho}(6)$
Jan	hist	14.8	242.1	0.636	0.342	0.236	0.193	0.164	0.130
	fitted	14.8	242.1	0.650	0.360	0.234	0.179	0.154	0.141
Feb	hist	13.4	228.7	0.756	0.508	0.392	0.293	0.204	0.132
	fitted	13.4	228.7	0.790	0.550	0.385	0.272	0.194	0.140
Mar	hist	12.6	200.2	0.621	0.367	0.269	0.229	0.165	0.107
	fitted	12.6	200.2	0.700	0.416	0.267	0.187	0.145	0.121
Apr	hist	8.47	83.9	0.725	0.552	0.476	0.402	0.371	0.330
	fitted	8.474	83.9	0.771	0.561	0.453	0.396	0.364	0.344
May	hist	5.384	37.6	0.650	0.472	0.428	0.467	0.413	0.334
	fitted	5.384	37.6	0.663	0.471	0.422	0.404	0.393	0.383
Jun	hist	3.569	11.6	0.638	0.451	0.360	0.296	0.279	0.267
	fitted	3.569	11.6	0.686	0.440	0.342	0.301	0.281	0.269
Jul	hist	3.243	21.9	0.629	0.297	0.218	0.159	0.150	0.182
	fitted	3.244	21.8	0.593	0.304	0.206	0.172	0.160	0.155
Aug	hist	4.398	77.0	0.544	0.234	0.164	0.137	0.114	0.105
	fitted	4.398	77.0	0.547	0.247	0.156	0.126	0.116	0.111
Sep	hist	4.351	46.4	0.504	0.340	0.294	0.231	0.195	0.208
	fitted	4.351	46.4	0.578	0.320	0.248	0.226	0.217	0.210
Oct	hist	7.423	125.8	0.656	0.416	0.318	0.277	0.301	0.297
	fitted	7.427	125.5	0.646	0.405	0.327	0.300	0.288	0.281
Nov	hist	11.04	222.9	0.659	0.457	0.383	0.336	0.295	0.271
	fitted	11.04	223.0	0.702	0.463	0.363	0.319	0.297	0.284
Dec	hist	12.8	178.2	0.672	0.438	0.357	0.256	0.142	0.092
	fitted	12.8	178.2	0.756	0.491	0.321	0.213	0.143	0.099

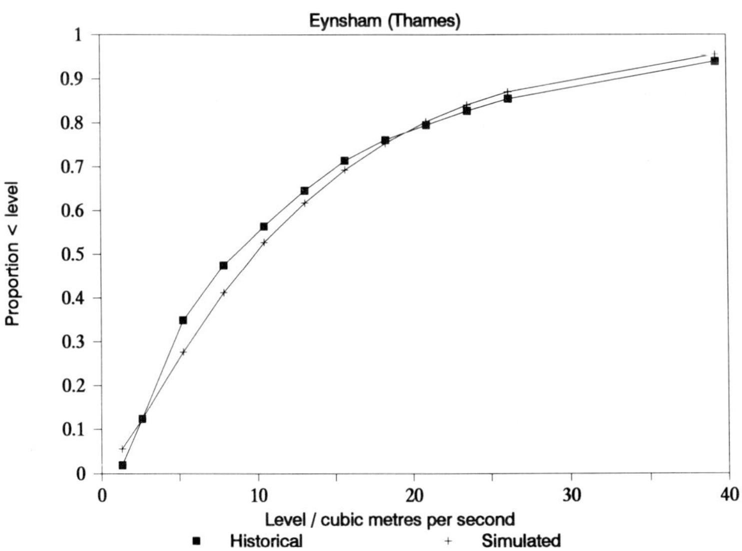

Fig. 4. Proportions of flow below selected levels for historical and synthetic flows: River Thames at Eynsham.

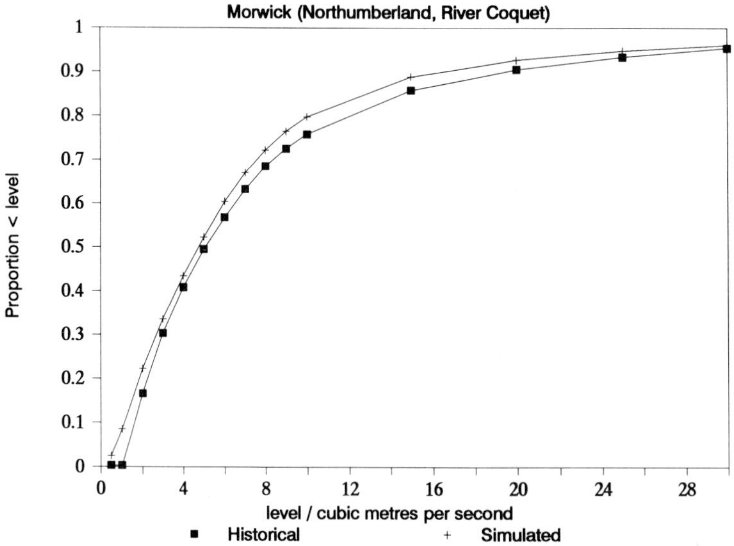

Fig. 5. Proportions of flow below selected levels for historical and synthetic flows: River Coquet at Morwick.

River Coquet (Figure 5). From these plots it is evident that the model shows a high degree of success in simulating the historical proportions of flow over the range chosen.

In addition, the maximum daily flow was found for each year of both the historical and simulated records. These maxima were then ordered and plotted against the standardised Gumbel variate (Figures 6 and 7). For the River Thames (Figure 6) it is evident that the model overestimates the extreme values. However, for the River Coquet the model matches the historical extremes quite closely (Figure 7). Hence, the model's fit to the historical extreme values appears to be site dependent. A better fit could be obtained by deriving the third moment of the NSSN model and using it in the fitting procedure.

To assess how well the model preserves the persistence present in the historical records, the historical and simulated monthly time series (i.e. average flows over months) were deseasonalised by subtracting the overall monthly means from the series. The correlograms were then plotted for these residual monthly series (Figures 8 and 9). From these plots it is evident that some of the historical persistence is being matched by the model. For example, in the case of the River Thames (Figure 8), the simulated correlogram matches closely the historical correlogram up to lag 3 but then decays more quickly thereafter. For the River Coquet, the simulated correlogram overestimates the historical lag 1 autocorrelation, matches the lag 2 value closely and again decays more rapidly (Figure 9). While there is evidence to suggest that the model can reproduce some of the observed persistence at the monthly level, this aspect of model performance could possibly be improved by lowering the upper bound on the parameter τ_2, or by adapting

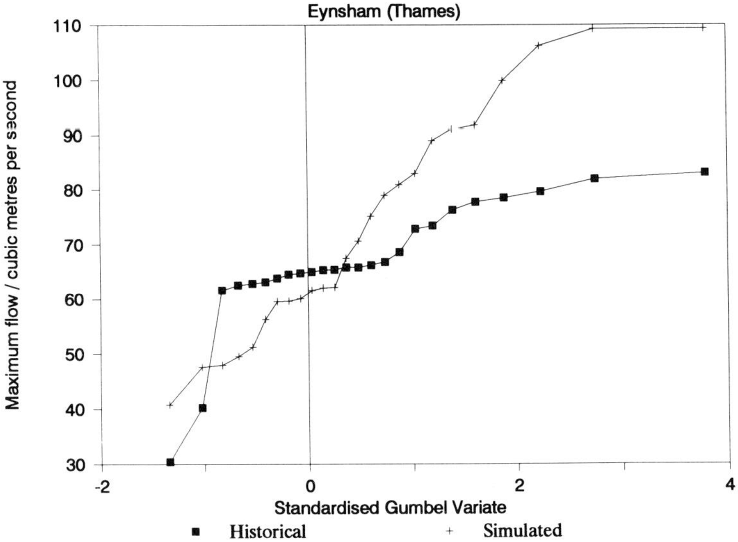

Fig. 6. Plot of annual maximum daily flows for historical and synthetic data: River Thames at Eynsham.

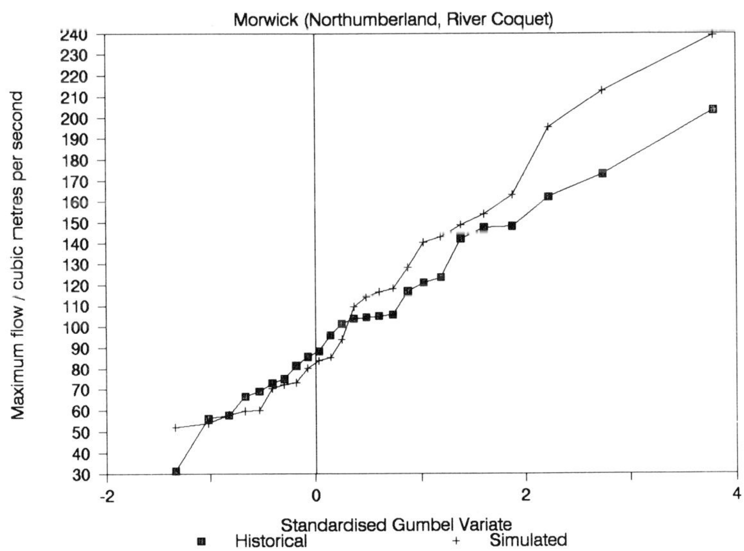

Fig. 7. Plot of annual maximum daily flows for historical and synthetic data: River Coquet at Morwick.

equation (9) so that the monthly autocorrelations could be used in the fitting procedure.

Finally, time series plots of the simulated data were studied to assess how well the model preserved the physical appearances of the historical plots [Figures 10 (Thames) and 11 (Coquet) are given as typical examples]. In comparing Figure 2

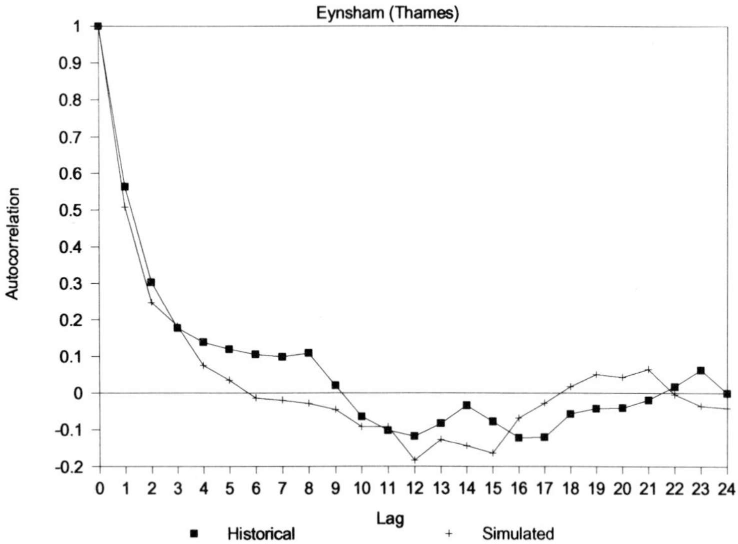

Fig. 8. Correlogram of standardized monthly historical and synthetic data: River Thames at Eynsham.

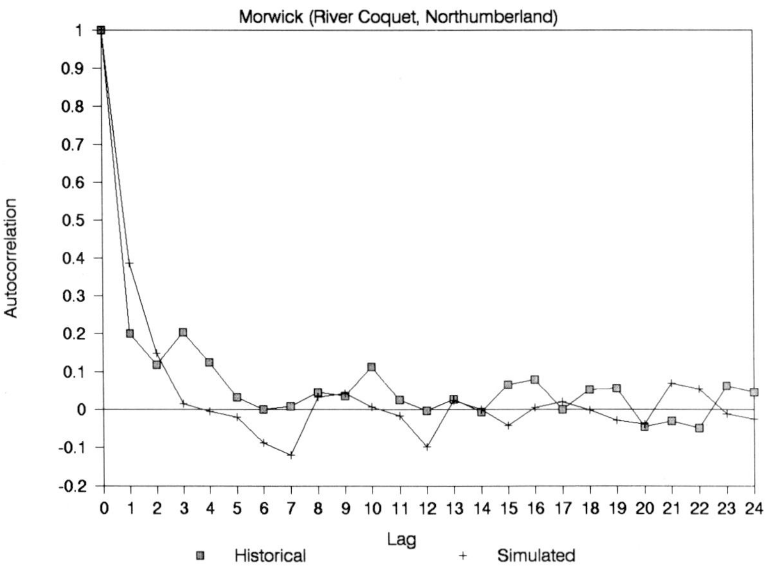

Fig. 9. Correlogram of standardized monthly historical and synthetic data: River Coquet at Morwick.

with Figure 9 and Figure 3 with Figure 10, it is evident that the main features of the historical plots are to be found in the simulated plots. However, for the Thames, there would appear to be too many summer peaks in the simulated data. This is likely to be due to the observed catchment flow from the winter months being mainly responsible for the flow level in the summer and not the rate of summer

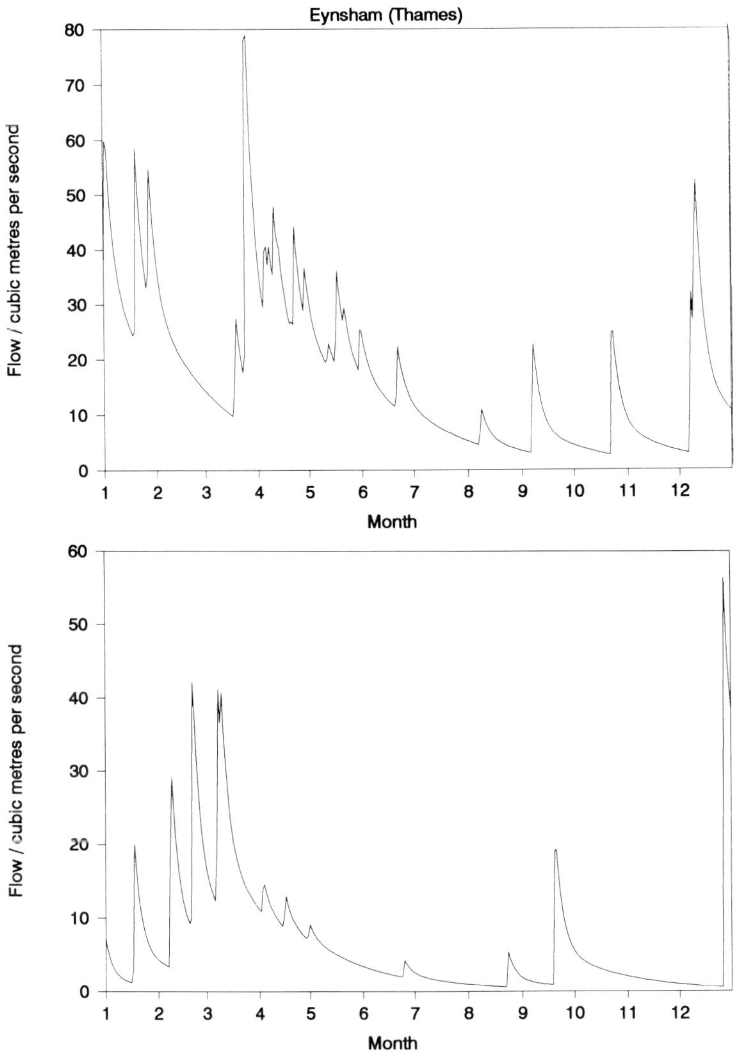

Fig. 10. Plots of two years of synthetic daily flows: River Thames at Eynsham.

storm arrivals (which is assumed when fitting the model). For the River Coquet there are no obvious discrepancies between the historical and simulated time series plots and the recessions are modelled quite satisfactorily.

7. Discussion and epilogue

While the assumption of a dependent structure for the event process in the NSSN model is plausible, the results obtained for the two data sets analysed would

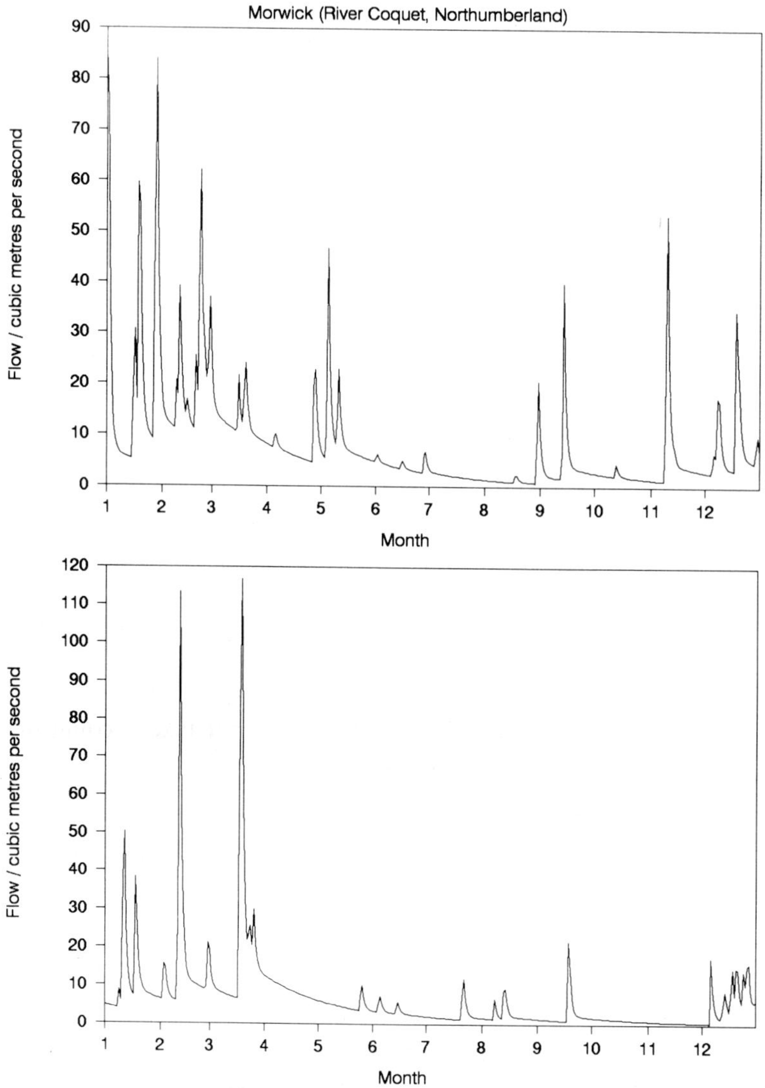

Fig. 11. Plots of two years of synthetic daily flows: River Coquet at Morwick.

suggest that some of the parameters of the event process cannot be satisfactorily resolved from daily data for a number of reasons. Firstly, a catchment smooths the small scale structure of rainfall, with the degree of smoothing increasing with catchment area and with the proportion of groundwater discharge in the catchment response. Secondly, averaging over a period of a day clearly masks any shorter term fluctuations in streamflow. Nonetheless, the additional parameters associated with a dependent event process can be expected to help in achieving a better fit to the historical statistics, as in the case of μ_c for the Thames (Table 2), even if their physical interpretation is not feasible. More physically realistic estimates of the event process parameters could possibly be obtained by using shorter-term (e.g. hourly) data but this will still not totally overcome the natural smoothing carried out by the catchment in continuous time. Another possibility would be to apply a Neyman-Scott model directly to short-term rainfall data and then use a rainfall runoff model to generate the effective rainfall values; these could then be used to parameterize the NSSN event parameters independently from the streamflow data. However, this would add considerably to the complexity of the whole exercise, and would bring with it all the attendant problems of building a satisfactory rainfall runoff model. Nonetheless, a simple runoff production function such as that used in the Xinanjiang/Arno model (Zhao et al., 1980; Dümenil and Todini, 1992) could prove to be a feasible option.

The parameter estimates obtained for the linear response component of the model are physically reasonable, particularly the larger proportions of groundwater runoff for the Thames in each month. However, in the summer months, the simulated baseflow for the Thames tends to decay more rapidly than in the historical flows. This could be improved by using monthly autocorrelations in the fitting procedure. However, the derivation of a modified form of equation (9) might not be straightforward, since adjacent months have different means and variances, thus invalidating the assumption of stationarity. An alternative would be to constrain the parameter τ_2 to induce greater carry over from the winter months. But this may lead to some distortion in the preservation of the means and variances in these months, since the NSSN model does not take account of flow contributions from preceding months when the parameters are estimated on a month by month basis. The fitting procedure could be modified to take account of this.

While the NSSN model has performed well in predicting the proportions of flow below certain levels, a more detailed assessment of its crossing properties is required before adopting it for practical use. In this context, the fitting procedure adopted offers considerable flexibility in fitting the model to reproduce those properties of primary concern for a particular application. But since the main theme of this volume is theoretical hydrology, this hopefully gives the authors some licence to set aside practical issues and extol the virtues of the theory. This brings to mind the theoretical hydrologist's prayer, often recited by Professor Dooge:

"Let all my systems be linear, and all my inputs Gaussian."

While the systems must indeed be linear, the inputs do not always have to be Gaussian!

Acknowledgements

The data for the River Coquet were supplied by the National Rivers Authority (Northumbrian Region), and for the River Thames by the Institute of Hydrology.

References

Box, G.E.P. and Jenkins G.M., 1970. Time Series Analysis Forecasting and Control. Holden Day, San Francisco, Calif., 553 pp.

Cowpertwait, P.S.P., 1991. Further developments of the Neyman-Scott clustered point process for modeling rainfall. Water Resour. Res., 27(7): 1431–1438.

Dooge, J.C.I., 1959. A general theory of the unit hydrograph, J. Geophys. Res., 64(2): 241–256.

Dümenil, L. and Todini, E., 1992. A rainfall runoff scheme for use in the Hamburg climate model. In: J.P.J. O'Kane (Editor), Advances in Theoretical Hydrology — A Tribute to James Dooge. Elsevier, Amsterdam, pp. 129–157 (this volume).

Lawrence, A.J. and Kottegoda, N.T., 1977. Stochastic modeling of riverflow time series. J. R. Stat. Soc., Ser. A, 140(1): 1–47.

Lungu, E.M. and Sefe, F.T.K., 1991. Stochastic analysis of monthly streamflows. J. Hydrol., 126: 171–182.

McLeod, A.I. and Hipel, K.W., 1978. Simulation procedures for Box-Jenkins models. Water Resour. Res., 14(5): 969–975.

Numerical Algorithms Group, 1977. Fortran Library Manual, NAG Executive, Oxford.

O'Connell, P.E., 1974. A simple stochastic modelling of Hurst's law. Proc. Int. Symp. on Mathematical models in Hydrology, Warsaw 1971, IAHS Publ., 100: 169–187.

O'Connell, P.E. and Jones, D.A., 1979. Some experience with the development of models for the stochastic simulation of daily flows. In: E.A. McBean et al. (Editors), Inputs for Risk Analysis in Water Resources. Fort Collins, Colorado, Water Resources Publ., pp. 287–313.

Rodriguez-Iturbe, I., Gupta, V.K. and Waymire, E., 1984. Scale consideration in the modelling of temporal rainfall. Water Resour. Res., 20(11): 1611–1619.

Rodriguez-Iturbe, I., Cox, D.R. and Isham, V., 1987. Some models for rainfall based on stochastic point processes. Proc. R. Soc. London, 410: 269–288.

Salas, J.D. and Obeysekera, 1982. ARMA model identification of hydrological time series. Water Resour. Res., 18(4): 1011–1021.

Trenberth, K.E., 1984. Some effects of finite sample size and persistence on meteorological statistics. Part 1: Autocorrelations. Mon. Weather Rev., 112.

Vandewiele, G.L. and Dom, A., 1989. A non-Gaussian multicomponent model for river flow. Water Resour. Res., 25(3): 397–404.

Weiss, G., 1977. Shot noise models for the generation of synthetic streamflow data. Water Resour. Res., 13(1): 101–108.

Zhao, R.-J., Zuang, Y.-L., Fang, L.-R., Liu, X.-R. and Zhang, Q.-S., 1980. The Xinanjiang Model. Proc. Symp. on Hydrological Forecasting, Oxford, IAHS Publ., 129: 351–356.

Part B. Large-scale hydrology

Chapter 7

Criteria for a hydrologically sound structuring of large scale land surface process models

ALFRED BECKER

Abstract

The assessment of land surface heterogeneity in large scale modelling of moisture and energy fluxes at the soil–vegetation–atmosphere interface is one of the primary tasks in recent world climate and global change modelling. After a brief characterization of scale ranges applied in hydrology and of their relation to different categories of hydrological models, the application of lumped models and the extrapolation of the validity of microscale process descriptions to the macroscale are criticized with reference to selected examples of heterogeneous land surfaces. Appropriate alternative modelling principles which take into account the areal pattern of atmospheric forcing as well as landscape patchiness and intra-patch heterogeneity are then briefly described and a hierarchy for areal discretization in land surface modelling as well as criteria to delineate zones of "uniform" atmospheric forcing are introduced.

1. Introduction

The world is faced with a massive and increasing human intervention in natural systems on a global scale, with consequences which, in several places, have reached a critical level and have begun to affect the world's climate and the living conditions on the whole globe. Predictions are needed in order to understand what future developments may take place and what measures need to be taken to avoid or reduce negative consequences for the welfare of humanity.

Remarkable progress has already been achieved in modelling the world's climate. However, what we really need is a hypothetical planet which, following Shuttleworth (1991), may be called "Modellion". It should start from the simple and evolve in time to become more like the Earth, including existing General Atmospheric Circulation Models (GCMs), ocean models and land surface models.

This paper relates to the modelling of the land surface of the Earth. The state-of-the-art, the given problems, shortcomings and perspectives in this modelling are comprehensively discussed in the above mentioned publication by Shuttleworth (1991).

One of the main shortcomings of the land surface component in existing GCMs is their inadequate spatial resolution. The spatial resolution of GCMs is about 10^4 to 10^5 km^2, which is incompatible with the characteristic scales of surface hydrologic processes. What cannot be explicitly resolved within the numerical model discretization must be parameterized.

Most of the parameterizations used in the past for describing land surface processes in GCMs are termed "lumped", e.g. "big leaf-big stoma" representations, which imply that land is homogeneously covered by a single big leaf within a grid element of the numerical climate model. This big leaf usually has a single stoma which is sensitive, in the most sophisticated parameterizations, to the environmental conditions known to have an effect on the mechanism of the stomata (i.e. solar radiation, temperature, humidity, carbon dioxide and soil water potential in the root zone). This stoma controls the plant transpiration and, as a result, the Bowen ratio and the surface heat fluxes over the GCM grid element as a whole (Avissar, 1991).

It is evident that this approach, in view of the natural variability of the landscape and the non-linear nature of the processes involved, is conceptually inappropriate and needs to be improved.

The question is how can this be achieved, i.e. how can the significant areal heterogeneities of larger land surface areas, such as GCM grid or sub-grid elements (about 100 to 10^5 km^2), or river basins (basin areas from about 10 to several 1000 km^2), be assessed appropriately in the modelling process at larger scales, particularly at the macroscale, see for example Dooge (1985) and Kühnel et al. (1991).

The aim of this paper is to contribute to this task, to discuss appropriate areal discretization schemes and to provide some guidance for their application in macroscale hydrological modelling. First, the definition of scales and the validity of available hydrological models in relation to the different scales, will be discussed briefly.

2. Scales in hydrology and related categories of hydrological models

It is widely agreed that the range of scales in hydrology is not equivalent to that used in atmospheric sciences. They are generally one order of magnitude smaller (e.g. Becker and Nemec, 1987). A more recent definition of scales in hydrology which takes into account the ranges of different categories of hydrological models is given in Table 1.

Two "transition ranges" (from micro- to mesoscale and from meso- to macroscale) are also defined in Table 1. They may be called the meso-γ and meso-α ranges, by analogy with the definition of scale used in atmospheric sciences.

Concerning the relation of model categories to hydrological scales, Dooge (1985) has written:

> [Large scale hydrological modelling usually means that] "... the finer scale processes may be either ignored or may be represented by their statistical

TABLE 1

Definition of scales in hydrology

Scales	Characteristics	
	Length (km)	Area (km^2)
Macroscale	$>10^2$	$>10^4$
Transition range (meso α)	$30–10^2$	$10^3–10^4$
Mesoscale (β)	3–30	$10–10^3$
Transition range (meso γ)	$10^{-2}–3$	$10^{-4}–10$
Microscale	$<10^{-2}$	$<10^{-4}$

effects in the large scale description. ... It has been found in practice that the models based on continuum mechanics are too complex to allow for the spatially variable nature of hydrologic systems to be taken into account in large scale modelling and they have been simplified to such an extent that they became in effect simple conceptual models."

These statements are essential in two respects:

(1) They indicate that the fundamental differential equations of continuum hydro- and thermodynamics (category 1 models) seem to be inappropriate for large scale modelling of land surface processes, due to the significant spatial variability of land surface characteristics (soil, vegetation, topography, etc.) and to the related fact that the conditions of internal homogeneity and continuity are often not fulfilled for these processes at larger scales.

(2) A different category of model, called "conceptual" (category 2), is mentioned as more appropriate for applications at larger scales (meso- and macroscale).

It should be said that both statements do not apply to large closed (continuous) water bodies, such as lakes, large rivers, aquifers, etc.

There are several well-developed and widely-applied conceptual hydrological models which can be classified as category 2 models (e.g. Becker and Serban, 1990). Some of them are physically based, in the sense that they were derived using acceptable simplifications, from the equations mentioned in (1), or they represent appropriate analogues of physically sound models, for instance diffusion type models, storage reservoirs, translation elements, cascades of such elements, etc.

Experience in applying these models indicates that they are most suitable for areas ranging from about 10 km^2 up to a thousand or a few thousand km^2 (about 3×3 up to 30×30 km). Considerable difficulties and losses in simulation quality occurred, however, when the same models and approaches were applied to smaller or larger areas.

Accordingly it may be concluded:

(a) Conceptual physically based models (category 2) are most appropriate for modelling of land surface processes at the hydrological mesoscale.

(b) In the macroscale range, the model area may be subdivided into sub-areas of the mesoscale size and category 2 models may then be applied to these sub-areas.

(c) Modelling at the microscale can be based on the fundamental differential equations of hydro- and thermodynamics with a detailed spatial resolution.

Special comments on the application of category 2 models will be made in a later section.

3. Critical concepts in macroscale hydrological modelling at land surfaces

The following two approaches are considered to be hydrologically critical in large scale land surface modelling:

(1) the "lumped" modelling approach, with averaging of process characteristics (forcing inputs, state conditions, land surface properties, etc.) over large areas (e.g. GCM grid areas of 10^4 to 10^5 km^2);

(2) the extrapolation of the validity of microscale process descriptions from the micro-scale (patches, vegetation stands, etc.) to the macroscale.

Two examples should illustrate the inadequacy of approach (a) from the hydrological point of view.

Example 1

A GCM grid square covers a predominantly flat and hilly terrain (70% of area) where the long-term average annual precipitation P ranges from 300 to 500 mm (areal average 400 mm) and the corresponding evapotranspiration ET ranges from 280 to 420 mm (areal average 350 mm).

On one side of the grid a mountain range is located (30% of the grid area) where P is significantly higher, up to 1500 mm, and ET increases to 530 mm, due to the availability of more water.

The related areal averages are listed in the following table, together with the long-term areal average runoff ($R = P - ET$):

(mm/a)	Lowland part (70% of area)	Mountain part (30%)	Whole grid area (100%)
P	400	1000	580
ET	350	460	383
R	50	540	197

By integration over the whole grid area the amounts listed in the right column are calculated:

$$Y \text{ (whole grid)} = 0.7 \times Y \text{ (lowland)} + 0.3 \times Y \text{ (mountains)}$$

As can be seen in the table, the 30% mountainous area of the GCM grid square produces nearly 83% of the total runoff, while the larger lowland part produces only 17%. It is evident that precipitation, as the primary forcing input to the system, is responsible for the significant difference in long-term average runoff between the mountain part and the lowland part of the grid unit, and accordingly, in the hydrological behaviour in general. If this difference is not taken into account explicitly, for instance, in separate models for the two sub-areas, the physical soundness of the process description is automatically lost and parameterizations become "arbitrary", even if the model has proved to be "well-fitted" in the "calibration" period.

Example 2

Another critical case is that of some oases within a desert or semi-desert. The oases, including the surrounding vegetated area in contact with groundwater, may cover 2% of the reference area (e.g. a GCM grid square). For simplification it may be assumed that no transition zone exists between the oases and the desert which varies in size depending on groundwater levels. The following long-term averages of evapotranspiration amount may be given:

(mm/a)	Oases (2% of area)	Desert (98%)	Whole area of grid square (100%)
ET	2600	15	67

The evapotranspiration of the oases is identical with the potential rate. Although the partial area is only 2% of the reference area, it contributes with about 78% ($0.02 \times 2600 = 52$ mm) to the total grid square evapotranspiration. The rest, 15 mm, about 22%, originates from a few rainfall events in the desert.

The separate determination of *ET* for the two types of sub-areas is conceptually clear, physically sound and rather simple, and so is the areal integration over the total GCM grid unit. Any lumped model for the grid unit as a whole, however, "violates" reality because it tries to make the desert area "a bit wetter", so that it can produce an increased *ET*, which in fact originates from the oases. Hydrological soundness is clearly lost and model parameters become just "tuning variables."

Therefore the application of lumped models for large areas like GCM grid units as a whole, can only be tolerated until more suitable distributed or semi-distributed models become available, and as long as other shortcomings in GCMs are more critical (as e.g. flux corrections).

The other critical concept is the application of microscale models on large scales. This concerns, for instance, the application of infiltration models or Soil–Vegetation–Atmosphere–Transfer schemes (SVATs) like the "big leaf-big stoma" parameterization, which has already been discussed briefly in the introduction. This discussion is not extended here. It should only be complemented by the following statement of Fiering (1982):

> "One of the assumptions frequently made is that our understanding of the microscale elements and processes (in the hydrologic cycle) can, with minor modifications, be extrapolated in principle to the understanding of the

macroscale environment, thus enabling reliable predictions to be made by linking the solutions to form a causal chain. Unfortunately, it seldom happens that way. Sooner or later, at some scale or characteristic dimension, mechanistic explanation breaks down and is necessarily replaced by unverified causal hypotheses or statistical representations of the processes."

A large number of existing SVATs have recently been reviewed by Geyer and Jarvis (1991). This review includes comments on the scales of application.

The limitations of infiltration models from the point of view of large scale application, have been examined by Entekhabi and Eagleson (1991) who developed an improved land surface parameterization scheme, which takes into account areal distributions of precipitation and surface characteristics in a statistical way.

In summary it must be said that it is timely to overcome the shortcomings in macroscale hydrological modelling characterized above.

4. Hydrologically sound structuring of macroscale models

Klemes (1985) in particular, critically examined the transportability of existing hydrological models and their applicability for long-term predictions in different environments and for changing conditions. He concluded that models for predictions should fulfil the following general requirements:

(1) The model structure must have a sound physical foundation and it must be possible to validate each of the structural components separately.

(2) The models must be geographically transferable, and their parameters must be derivable from estimates of real-world characteristics at any place on the Earth.

These requirements are in agreement with what has been said already. The question is now what possibilities are there for adequate, physically sound, vertical and horizontal structuring of macroscale hydrological models.

Figure 1 illustrates the basic structure of macroscale hydrological models which consist of a water balance model and a transport model (Vörösmarty and Moore,

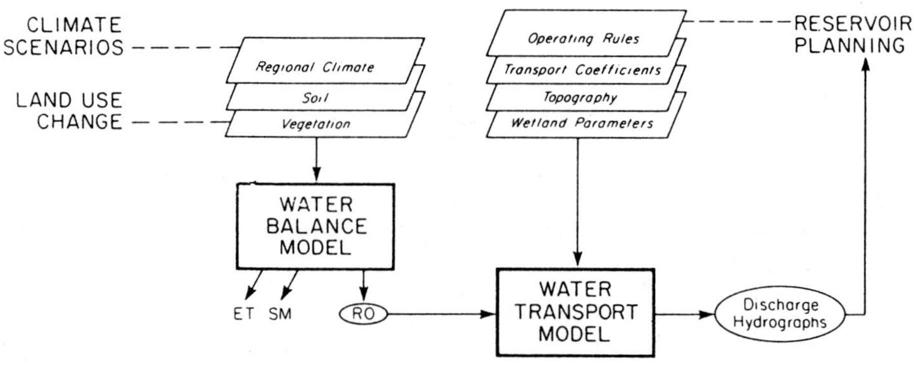

Fig. 1. General structure of macroscale hydrological models (from Vörösmarty and Moore, 1991).

1991). By definition the water balance model is more related to the domain of vertical energy and moisture fluxes at land surfaces, while the water transport model covers the domain of lateral flows (surface and sub-surface).

Modelling of the vertical fluxes can be related to any land surface area of interest (grid units, patches, river basins, etc.), while water transport models must be related principally to hydrological systems (river basins, groundwater systems, etc.). Taking this into account Becker and Nemec (1987) have suggested a so-called "two-level" modelling concept, in order to ensure that different areal discretization schemes can be applied to the two "levels" understood as the two domains defined before. In the following the term "two-domains modelling concept" is preferred.

Figure 2 illustrates the sub-processes of the hydrological cycle, and related sub-systems, that have to be considered in hydrological modelling and which sub-systems belong to the "vertical fluxes domain" or to the "lateral flows domain". The following facts are particularly important for distinguishing between the two domains of modelling:

(a) In the "vertical fluxes domain" any required or reasonable, and more or less detailed, areal discretization scheme can be applied:

- on regular grids of any size (GCM grids or sub-grids);
- with sub-areas of "equal" hydrological behaviour, with respect to land surface properties, forcing atmospheric inputs, etc. An a priori relationship to watersheds, river basin divides, etc., is not necessary.

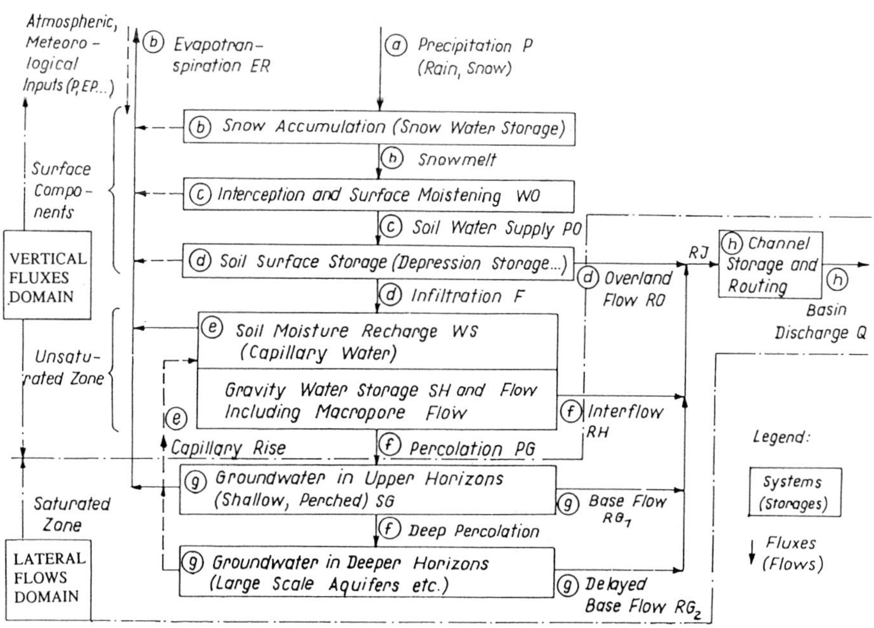

Fig. 2. Components and sub-processes of the land phase of the hydrological cycle.

(b) Modelling in the "lateral flows domain", however, must be related from the very beginning to watersheds, even in large scale regional and continental modelling. Component models for this domain are clearly watershed related; and in many cases simplified, physically based or conceptual models are sufficient for river basins or sub-basins as a whole (category 2 models).

(c) In river basin studies only those parts of the "vertical fluxes domain" model must be coupled to the "lateral flows domain" model which lie within the river basin.

For GCMs the "vertical fluxes domain" is particularly important. "Lateral flows domain" models are of interest only insofar as observed flows are intended to be

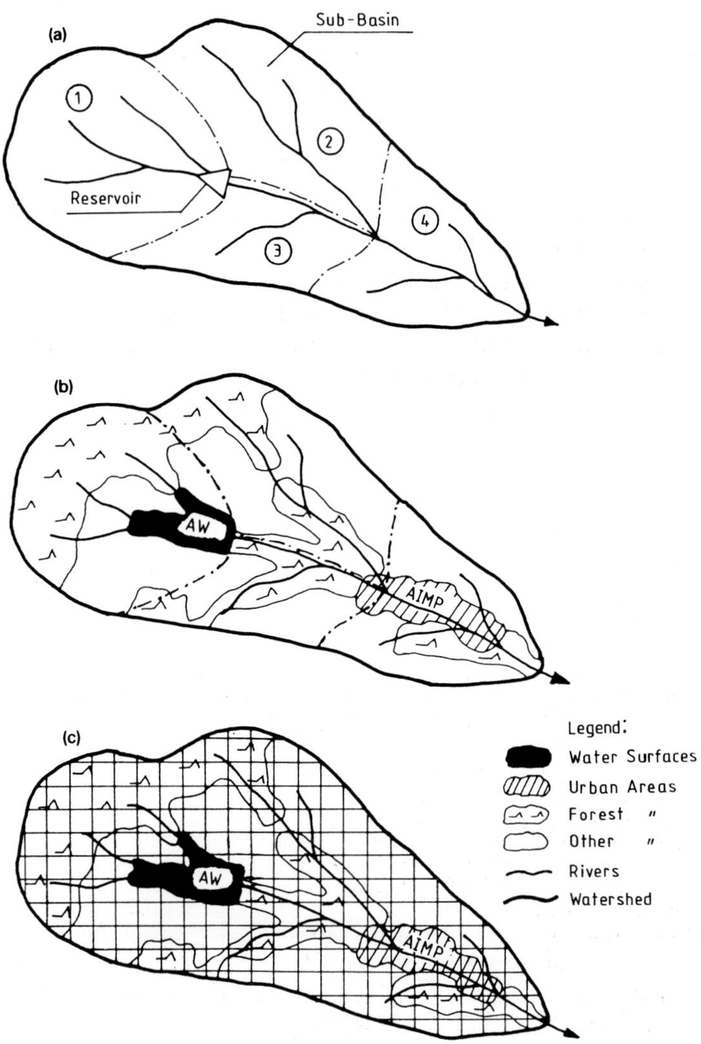

Fig. 3. Representation of different space discretization schemes in river-basin modelling: (a) lumped (4 sub-systems); (b) semi-distributed; (c) distributed (grid-based).

used for validation purposes, or for the investigation of interactions between water resources systems and climate.

The most essential task in modelling the "vertical fluxes domain" is what areal discretization scheme should be applied to guarantee a hydrologically sound assessment of areal heterogeneity and to avoid the misconception explained in section 3. The main alternatives are represented in Figure 3 (Becker and Serban, 1990).

The application of a high resolution fine meshed regular grid as represented in Figure 3c is always possible but troublesome, and not always efficient. Therefore the semi-distributed approach is considered as an appropriate compromise solution. It allows one to take into account, in an appropriate way, areal process differentiation as discussed in section 3. Sub-areas to be modelled separately, can be chosen as large as possible, and must not be split into small pieces.

5. Basic types of areal heterogeneity at land surfaces and assessment of landscape patchiness

There are two basic types of heterogeneities that need to be considered (Avissar, 1991): (a) landscape patchiness, and (b) intra-patch heterogeneity.

Landscape patchiness means the succession of different types of land surface areas: a mosaic of agricultural fields, forests, built-up areas, water surfaces etc. on various topographical, pedological, geological and other features. Patches generally represent hydrological units with an "approximately equal" or "uniform" hydrological behaviour.

Intra-patch heterogeneity refers to the small scale variability of soil, vegetation and other parameters as, for instance in a vegetated patch, root depth, stomatal resistance, etc. Intra-patch heterogeneity is generally more or less random in its distribution, so that statistical approaches are best suited for its assessment. The modelling of patches with clearly distinct hydrological regimes may require model components that are differently structured.

The first important question in assessing landscape patchiness is how many types of patches need to be taken into account.

Penman, who looked at the problem from the point of view of evapotranspiration, suggested two classes of land surface:

(a) Riparian areas where the groundwater table would be close enough to the surface for evaporation to be always at the potential rate (shallow groundwater areas with a capacity for exfiltration).

(b) Non-riparian areas where groundwater would be well below the surface and the actual rate of evapotranspiration during dry periods is solely dependent on the moisture content in the unsaturated zone (areas with deep groundwater with a capacity for infiltration).

This is a very important approach which immediately solves the "oasis problem" as described in section 3, as well as many similar problems in other landscapes.

TABLE 2

Patch types with significantly different hydrological regime

Patch type	Symbol	Evapotranspiration	Direct runoff
Water areas	AW	Potential	Equal to precipitation (100%)
Impervious areas (sealed, rocks)	AIMP	Only during precipitation events (i.e. generally small)	Equal to precipitation (close to 100%)
Flat areas with:			
Deep groundwater (infiltration areas)	AF	Depending on the soil moisture in the root zone	Generally close to zero
Shallow groundwater (exfiltration areas)	AN	Generally potential	With increasing water storage equal to precipitation at saturated sub-areas
Hillslopes with:			
Deep permeable underground layers (infiltration areas)	AG	As AF above	As AF above, but infiltration excess possible during heavy rainfall or with frozen ground
Shallow soils above less permeable layers or bedrock	AH	As AF above	From saturated sub-areas equal to precipitation, otherwise infiltration, excess and interflow

Subsequently, when runoff formation was the object of deeper research, hillslope areas and built-up areas with very high percentage of impervious surface were taken into account as additional patch types in order to improve the simulation of direct runoff (overland flow and interflow). A review of this development is given by Kühnel et al. (1991).

A more recent classification of patch types is given in Table 2 (Becker and Pfützner, 1986). It combines consideration of both evapotranspiration and runoff formation and represents an appropriate compromise towards a hydrologically-sound discretization of land surface areas. Of course, other patch types (forests, pasture, different crop stands, ...) can also be taken into account, if this is essential.

In any case, the decision on how many types of patches to consider, should be made carefully. It depends strongly on the problem to be solved. In macroscale modelling it will often be sufficient to distinguish between only two evapotranspiration related types, (AW + AN) in Table 2 on the one hand, and all other types on the other. It may be more appropriate to consider four types: AW, AIMP, (AF + AG + AH) and AN.

When dealing with the components of runoff, the six types of Table 2 should be distinguished. For more detailed biosphere studies, consideration of different crop stands may also be required.

After the decision has been made, which patch types to take into account, the subdividing (disaggregation) of the model area into sub-areas can be done (on maps, air photographs etc.). An important simplification can then be achieved, by combining all single patches of a special type within the model area into one arbitrary unit. This unit can then be modelled with an adequate component model.

If the accumulated total area of special patch type i within a model area (e.g. grid element) is defined as A_i, then the total flux of a certain type k (sensible heat, latent heat, long wave radiation or others) from the whole model area is given as:

$$F^{(k)} = \frac{\sum_{i=1}^{n} A_i \cdot f_i^{(k)}}{\sum_{i=1}^{n} A_i} \tag{1}$$

with $f_i^{(k)}$ the particular flux of type k from patch type i. The number of patch types considered is n.

This approach clearly represents a "semi-distributed" approach according to Knudsen et al. (1986) and Becker and Serban (1990). It neglects the real distribution of the different individual patches within the model area and their interaction with neighbouring patches of different types (advection effects etc.), but it overcomes the most critical effects of averaging the different hydrological behaviours of different patch types over large areas.

There are several examples of successful application of this modelling technique to river basins (e.g. Knudsen et al., 1986; Becker and Pfützner, 1986; Avissar and Pielke, 1989; Kaden et al., 1991). They cannot be described in the framework of this paper. However, in summary it can be said, that the appropriateness, usefulness and efficiency of the semi-distributed approach for large scale modelling has clearly been confirmed. The only question that remains is the discovery of the conditions under which this simplification can be accepted.

Intra-patch heterogeneity is usually random and occurs on small scales (microscale). Accordingly, it is more appropriate to assess this heterogeneity by means of statistical approaches, e.g. by applying distribution functions for the relevant land surface and other characteristics in a model of the study area such as a patch. Several such approaches have been developed and are described elsewhere. They concern the following processes and properties:

- water holding capacity of the rooted soil layer (Becker, 1975);
- runoff ratio, soil evaporation and transpiration (Entekhabi and Eagleson, 1989);
- different soil and plant characteristics (Avissar, 1991);
- topography, soil and plant characteristics (Famiglietti and Wood, 1991).

All these approaches are steps in the desired direction. However, they are generally applied without considering the constraint that their application is limited to areas (zones) of uniform atmospheric forcing, as discussed in the following sections and in example 1 in section 3. In any case, further investigation is needed.

6. Hierarchy in the areal discretization of land surfaces for modelling

In conclusion, a hierarchical approach is recommended in discretizing (disaggregating) land surfaces for hydrological modelling:

(1) Investigation of the areal pattern of essential characteristics (for instance long-term mean) of atmospheric driving forces operative on hydrological processes at land surfaces; in particular precipitation, radiation and related characteristics.

(2) Definition of ranges of these characteristics within which areal averaging is acceptable, at least in the long-term, and statistical distribution functions can be used to describe event specific areal differentiation (patterns); This has been done for instance by Entekhabi and Eagleson (1991).

(3) Subdividing (zoning) the study area into such sub-areas (zones) of "uniform" atmospheric forcing (in terms of long-term characteristics, in particular long-term means).

This zoning is essential and is needed to define larger areas which, from the point of view of atmospheric forcing, can be handled as units in large scale modelling. It avoids the problems explained in example 1 in section 3. Additional explanation will be given in the following section. An example of zoning is given in Figure 4, left side.

(4) Definition of patch types to be modelled separately, in consideration with the remarks made in the previous section.

(5) Assessment of landscape patchiness, i.e. sub-division of the study area into patches and determination of the areal percentage of each patch type considered in each zone of uniform atmospheric forcing (aggregation of individual patches). An example is given in the right side of Figure 4.

(6) Selection of an appropriate model for each patch type which may or may not include elements for assessing intra-patch heterogeneity.

(7) Estimation of the parameters of the selected patch type model.

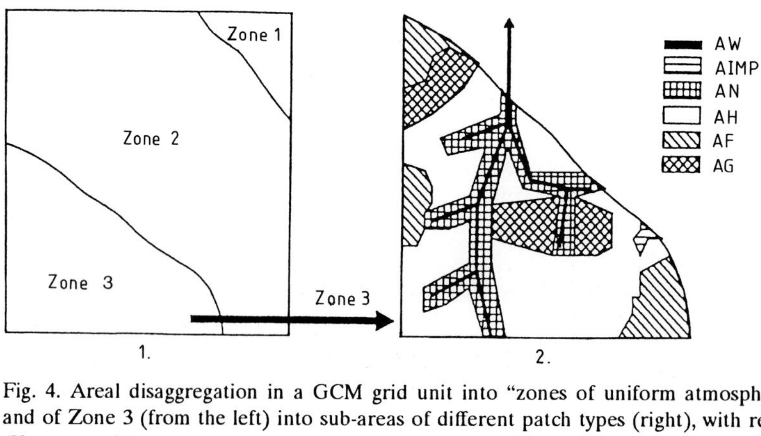

Fig. 4. Areal disaggregation in a GCM grid unit into "zones of uniform atmospheric forcing" (left) and of Zone 3 (from the left) into sub-areas of different patch types (right), with reference to Table 2 (Kaden and Pfützner, 1991).

(8) Composing the integrated model for the "vertical fluxes domain" as illustrated in Figure 2, by aggregating the component models of the sub-systems involved.

The "vertical fluxes" domain model allows one to describe and study all important hydrological processes at the land–atmosphere-interface. The coupling of models to it for the "lateral flows domain" is a separate modelling step related to the aquifer and river basin, as has been discussed in section 4.

Modelling of this domain is required if:

- the hydrological cycle must be closed downstream as far as the ocean;
- "vertical fluxes domain" models are intended to be validated with reference to measured discharges (records);
- water resources impact studies are planned;
- interaction between climate and water resources are to be investigated;
- problems of material transport in rivers, of nutrient cycling etc. are to be solved.

It should be pointed out again that it is the two-domains modelling-concept, briefly explained in section 4, which allows one to model separately both domains with different areal discretization schemes and to couple both models finally with reference to river basin divides (Becker and Nemec, 1987).

It is not possible in the framework of this paper to discuss further aspects of modelling the "lateral flows domain" and its coupling to the "vertical fluxes domain" model. This will be done in another paper.

7. Criteria to delineate zones of uniform climate forcing

As remarked upon in the previous section, the zoning of large land surface areas (e.g. GCM grid units) into "zones of uniform atmospheric forcing" is of particular significance. However, no general methodology is available for this purpose. Therefore, a proposal is made here which could contribute to such a methodology.

Let the local state or flux variable at instant t at any point x,y in a study area be:

$$u(x,y,t)$$

The long-term total of this variable is:

$$\overline{u}(x,y) = \int_{t_0}^{t_0+T} u(x,y,t)\,\mathrm{d}t \qquad (2)$$

where T is the integration period, e.g. 1 year, several years...

A zone of uniform climate forcing ACF may then be defined as an area where the local value of $\overline{U}$ does not deviate significantly from the areal average $\overline{U}_{ACF}$:

$$\overline{U}_{ACF} = \iint_{ACF} \overline{u}(x,y)\,\mathrm{d}x\,\mathrm{d}y \qquad (3)$$

If a tolerance deviation Δu is defined, the following non-exceedance criterion can be formulated for the definition of a zone ACF:

$$\left|\overline{u}(x,y) - \overline{U}_{ACF}\right| < \Delta u \tag{4}$$

u may be defined as a percent of $\overline{U}_{ACF}$, for instance 10 or 15%.

The above criterion is related to long-term areal totals or means. There may be other criteria which can equally be used, for instance variances, differences in seasonal variation patterns etc.

However, precipitation is particularly important for hydrology, natural resources and the environment, in particular areas or in a river basin in general. Long-term totals or averages of precipitation and related parameters determine the water balance in these areas. Therefore, the above criterion will be important. Further studies in this field are indispensable.

References

Avissar, R., 1991. A statistical-dynamical approach to parameterize subgrid-scale land surface heterogeneity in climate models. Surv. Geophys., 12: 155–178.

Avissar, R. and Pielke, R.A., 1989. A parameterization of heterogeneous land surface for atmospheric numerical models and its impact on regional meteorology. Mon. Weather Rev., 117: 2113–2136.

Becker, A., 1975. The integrated hydrological catchment model EGMO. In: Int. Symp. and Workshops on Applied Mathematical Models in Hydrology and Water-Resources Systems, Bratislava. Hydrol. Sci. Bull., 21: 1.

Becker, A. and Nemec, J., 1987. Macroscale hydrologic models in support to climate research. In: The Influence of Climate Change and Climatic Variability on the Hydrologic Regime and Water Resources. Proc. Symp., Vancouver, August 1987. IAHS Publ. No. 168, pp. 431–445.

Becker, A. and Pfützner, B., 1986. Identification and modelling of river flow reductions caused by evapotranspiration losses from shallow groundwater areas. 2nd Scientific Assembly of the IAHS, Symp. 52, Budapest, July 1986. IAHS Publ. No. 156, pp. 301–311.

Becker, A. and Serban, P., 1990. Hydrological models for water-resources system Design and Operation. Oper. Hydrol. Rep., No. 34, WMO-No. 740, Geneva.

Dooge, J.C., 1985. Hydrological modelling and the parametric formulation of hydrological processes on a large scale. WCP-Publ. Ser. No. 96, WMO/TD-No. 43, Geneva.

Entekhabi, D. and Eagleson, P.S., 1989. Land surface hydrology parameterization for atmospheric general circulation models including subgrid scale spatial variability. J. Climate, 2(8): 816–831.

Entekhabi, D. and Eagleson, P., 1991. Climate and the equilibrium state of land surface hydrology parameterizations. Surv. Geophys., 12: 205–220.

Famiglietti, J.S. and Wood, E.F., 1991. Evapotransipation and runoff from large land areas: Land surface hydrology for atmospheric general circulation models. Surv. Geophys., 12: 179–204.

Fiering, M., 1982. Overview and recommendations. In: Scientific Basis of Water Resources Management. National Academy Press, Washington, D.C.

Geyer, B. and Jarvis, P., 1991. A review of models of Soil–Vegetation–Atmosphere–Transfer-Systems (SVATS). TIGER III, University of Edinburgh, EH9 3JU, March 1991.

Kaden, S. and Pfützner, B., 1991. Large-scale modelling of land surface–atmosphere interface processes — Macrohydrologic software system. Report to UNESCO-IHP, Paris (SC/RP 205084.1).

Klemes, V., 1985. Sensitivity of water resource systems to climate variations. WCP Report No. 98, WMO, Geneva.

Knudsen, J., Thomsen, A. and Refsgaard, J.C., 1986. WATBAL — a semi-distributed, physically based hydrological modelling system. Nord. Hydrol., 17: 415.

Kühnel, V., Dooge, J.C.I., O'Kane, J.P.I. and Romanowicz, R.J., 1991. Partial analysis applied to scale problems in surface moisture fluxes. Surv. Geophys., 12: 221–247.

Shuttleworth, W.J., 1991. Modellion concept. Rev. Geophys., 29(4): 585–606.

Vörösmarty, C.J. and Moore, B. III, 1991. Modelling basin-scale hydrology in support of physical climate and global biogeochemical studies: An example using the Zambezi River. Surv. Geophys., 12: 271–311.

Chapter 8

The construction of continental scale models of the terrestrial hydrological cycle: an analysis of the state of-the-art and future prospects

L.S. KUCHMENT

Abstract

In this paper an attempt is made to discuss the possible approaches to modelling the hydrological cycle at the continental scale, taking into account the state-of-the-art in the relevant hydrological sciences, the availability of information, and progress made in the modelling of climatic systems.

1. Introduction

The need for models of the terrestrial hydrological cycle which can simulate, with sufficiently detailed resolution, the vertical and horizontal processes of water transfer on a global scale, is becoming more and more obvious.

A number of studies have already shown a significant influence of soil moisture and evapotranspiration on the continental distribution of precipitation and air temperature. This means that these land surface processes must be taken into account in the global climate system, even for short-term weather forecasting. On longer intervals of time, different distributions of soil moisture can influence river inflow to seas and oceans, and consequently, their salinity and the circulation of the deep ocean. Of great concern is the problem of possible impacts on climate of man-induced large-scale transformations of the hydrological cycle, caused by agriculture and changing land-use and runoff-control. Finally, the question arises concerning how global or large-scale changes, and first and foremost among these the change of climate, can affect hydrological systems in different parts of the Earth.

It seems that proper hydrological models, together with present-day large-scale models of meteorological processes, can, in the near future, help to (1) study a number of significant hydrological anomalies of continental scale, (2) confirm or reject different hypotheses in the explanation of these phenomena, and (3) predict their future evolution.

Among the most interesting large-scale hydrological events are: (1) the rise in the level of the Caspian Sea, which began in 1978 and has continued to the present, (2) the desertification of the Sahel and the drying of Lake Chad, (3) the hydrological consequences of deforestation in Amazonia, (4) the change of scale and frequency of flooding in the Ganges basin, etc.

Because all these problems have much in common, their research does not appear to be divisible into separate independent questions. All of them should be considered together, in modelling the climate system as a whole. To save money and labour, it is also desirable that the large-scale hydrological models, at different levels of aggregation, incorporate or use models of hydrological processes at regional or large river-basin scale, which include (a) river runoff formation, (b) active water resource systems, and (c) water transfers in large ecological systems: large forest territories, swamp areas, agricultural regions etc. In the past, such models have been developed at different temporal and spatial scales for many separate tasks: short-term and long-term hydrological forecasting, design and control of water resource systems, irrigation, reclamation, etc. To combine these efforts efficiently in future work, it is important to find ways of (a) choosing acceptable temporal and spatial scales, and (b) maintaining the correspondence between the physical meaning of coefficients and constants in different models.

The greatest difficulty in constructing large-scale models of the hydrological cycle is caused by the patch-type variations in land-surface properties and the shortage of information on these variations. This leads to the following important problems:

(1) the adequate assimilation of all available observational data, which can be used for the determination of land-surface characteristics influencing water transfer,

(2) the aggregation of these characteristics at the chosen time–space scales, and

(3) the taking into account of the small-scale variability of land-surface properties and hydrometeorological values in terms of the chosen scale: the sub-grid-scale effects in numerical models.

The construction of large-scale models of the hydrological cycle can be based on the experience acquired in developing runoff-formation models of large rivers; but, to guarantee sufficient accuracy in the description of (a) all the components of the hydrological cycle, and not only runoff, and (b) the interaction between hydrological and meteorological processes, and perhaps other processes of the bio-geosphere, such models cannot be a simple mechanical combination of the runoff models of many adjacent river basins.

2. Time space scales and main research directions

The basis of present-day methods of climate simulation is found in three-dimensional general circulation models (GCMs) which describe not only the global transfer of air mass and energy but also the main physical processes in the atmosphere: radiation balance, formation of cloudiness and precipitation, physics of boundary layer of atmosphere, land surface processes, etc. Currently GCMs are being steadily improved by adding more detailed descriptions of different processes and by using finer resolutions for numerical calculations. However, in spite of the significant progress achieved in modelling the global distributions of temperature and precipitation, the regional climate simulations produced by GCMs have insufficient spatial resolution and accuracy for hydrological use (e.g. in climate change impact studies). This can be explained, not only by the too-coarse spatial resolution of present-day GCMs, but also by the lack of a number of physical processes, which have a significant influence on meso-scale climate conditions and on the hydrological cycle. Most of the GCMs have horizontal resolutions in the range 2.5–10° (approximately 250 to 1,000 km). Such a grid does not make it possible to reproduce important topographic features (for example, large mountain systems) and large water bodies (for example, the Great Lakes or Aral Sea). Only recently have papers appeared on GCMs, where the horizontal resolution is about 1° (Mahlman and Umscheid, 1987) but this has been achieved by simplifying the description of a number of important processes.

At the present time, the main reason preventing GCMs from being run at meso-scale resolution (less than 100 km) is computing time. Giorgy and Mearns (1990) describe the results of numerical experiments with a version of the Community Climate Model of the National Center for Atmospheric Research in the USA carried out on a Cray X-MP computer at five different spatial resolutions (Table 1).

As is evident from Table 1, the computing time used to simulate one day of climate at a resolution of 300 km × 300 km is 5 minutes, at a resolution 200 km × 200 km, 15 minutes, but at a resolution of 30 km × 30 km, the computing time is more than 2 days. In this context, it is worth mentioning that, for many hydrological applications, it is desirable to simulate processes for dozens of years.

Another important reason why fine resolution GCMs cannot be realised in the near future, is the absence of proper meso-scale models of a number of important processes included in GCMs (e.g. cloud formation, precipitation, land processes). Most parameterisations of these processes (descriptions of the small-scale processes in terms of large-scale process variables) have been developed and fitted for coarse resolution models.

TABLE 1

Runs of GCM1 on the Cray X-MP computer: computer power needed for a given resolution

Resolution grid box size °Lat. × °Long. (km × km)	Number of horizontal grid points	Time step	Number of time steps per day	CPU * required for 1 day simulation
4.5° × 7.5° (500 km × 600 km)	1,920	1,800	48	1
2.9° × 2.9° (300 km × 300 km)	8,192	900	96	5
1.8° × 1.8° (200 km × 200 km)	18,432	600	144	15
1.1° × 1.1° (121 km × 121 km)	583,200	90	960	3,000

* CPU = Central Processing Unit.

However, much useful information for hydrological purposes can be derived from present-day GCMs. In many cases, empirical relationships can be established between large-scale and meso-scale meteorological characteristics, which can be applied to the simulation of the hydrological cycle in river basins. Some prospects for modelling regional climate and, hence, the hydrological cycle are offered by using a nesting procedure whereby a higher resolution model sub-region is embedded (nested) in a lower resolution global model (Giorgi and Mearns, 1990). In spite of the large amount of computing time involved, the application of nesting procedures in GCMs can be justified at least for the most hydrologically important sub-regions and time periods.

As for time scale, the time steps used, even in GCMs with fine resolution (15–30 minutes), are quite acceptable for hydrological tasks. In many cases it is possible to use larger time intervals.

Given the above analysis of the current state of GCMs and availability of relevant observational data, it is suggested that efforts in developing continental scale models of the hydrological cycle might be concentrated into the following three directions:

(1) The development and improvement of continental scale hydrological models and of parameterisations of the hydrological cycle in GCMs (including the sub-grid disaggregation of meteorological input data and the sub-grid aggregations of the outputs of hydrological models) for the present-day horizontal resolution of 100–300 km and for river basin areas in the range 10,000–90,000 km^2. A number of such models have already been developed, but they pay too little attention to runoff formation processes and to using runoff observation data. Models of this type can be used in present-day GCMs.

(2) The development of parameterisations of the hydrological cycle for GCMs with a horizontal resolution of 30–100 km (basin areas 1,000–10,000 km^2). On the basis of the present-day theory of hydrological processes, and taking into account the future potential of GCMs and expected developments in the collection of observational data, this development programme can be realised on a global scale within 5–10 years. It may be associated with a new generation of GCMs and the development of methods for the sub-grid disaggregation of meteorological data on finer spatial scales.

(3) A long-term programme, which concentrates on describing the hydrological cycle as a part of the climatic system for horizontal resolutions of less than 30 km. In addition to the problems of modelling and parameterising hydrological processes, such a programme must include additional theoretical and experimental studies of the physics of the hydrological cycle, taking into account small-scale variations of relief, soil and land use, as well as the development of remote sensing methods for measuring the characteristics of hydrological systems. The extent to which the aims of this programme can be realised, will depend on developments in computer science and space technology, notably the implementation of new methods of airborne and satellite measurement, and on continuing public interest in the problems of global change and weather prediction.

This division of research activity is conditioned by, and takes into account, the existing levels of present-day science and technology and the time–space scales which have to be used for the final calibration and application of hydrological models. There are many issues between these three research directions which must be addressed before they can be resolved. However, the preservation of a uniform description of processes for each of the above mentioned programmes, does not appear necessary. It seems likely that the construction of global models of the hydrological cycle will be carried out by different groups of scientists and from different regional blocks: Europe, several Asian blocks, South and North America, etc. But the delineation of these blocks must be based on river basin schematisations which take into account the grids of the GCMs. It is also desirable that continental scale hydrological models should arise as a result of the generalisation of models with smaller spatial resolutions.

3. Continental scale hydrological models with present-day space resolution of GCMs (100–300 km)

Hydrological models, which can be considered to be of continental scale, were developed many years ago: examples are the model of unsteady flow in the Missouri–Mississippi river system, the model of runoff formation in the Colombia basin, the model of flood routing in the Volga basin to Samara, the model of flood routing in the Mekong river. However, in these models, the descriptions of processes and the calibration of parameters have been conditioned by the main aim of achieving the desired accuracy in reproducing runoff hydrographs; the evapotranspiration and other components of the hydrological cycle have been described in simplified form or have not been considered at all.

On the other hand, in GCM representations of the terrestrial hydrological cycle, the main attention has been paid to describing the evapotranspiration element, and runoff has been commonly determined as the difference between precipitation and evapotranspiration. Only in the last decade, has it became clear that the incorporation into GCMs of the processes of runoff formation, the movement of soil moisture, snow cover formation and snowmelt, and other processes within the hydrological cycle can not only improve the calculation of some meteorological fields, but also be useful in the validation of GCMs as a whole, especially by including the processes of runoff formation on large areas and by using runoff observations data. The possibility of improving the description of the hydrological cycle in GCMs has also generated an increasing amount of interest in these models among hydrologists.

The description of hydrological processes in present-day GCMs is mainly limited to vertical moisture and heat exchanges between the land surface and the atmosphere; the horizontal transfer of water on the land surface and ground water are not usually described. One of the most important reasons for choosing a particular parameterisation of the hydrological cycle in a GCM, is to save computing time.

The simplest model of the hydrological cycle which has been implemented in a number of GCMs is the Budyko bucket model (Budyko, 1948). In this model the soil moisture content is increased by rainfall and snowmelt and is decreased by evaporation before the moisture content reaches a critical value: the "field capacity" which is usually taken as 15 cm). Any soil moisture in excess of the "field capacity" is assumed to form runoff. Evaporation occurs at the potential rate until the soil moisture reaches another critical value, usually 0.75 of the "field capacity". After this point the evaporation rate is assumed to be proportional to the ratio of the soil moisture content to "field capacity".

A large advantage of this model is that only two parameters are used, which are easy to fit. But it is obvious that to ignore the roles of vegetation, soil moisture movement and runoff formation is too crude an assumption. This was the impetus for developing more detailed parameterisations of vertical moisture exchange in GCMs.

The most detailed descriptions of these processes are used in the so called "big leaf" models. These models are based on the two assumptions: (1) the descriptions of moisture movement processes which occur at a point in the grid space can be transferred to large grid areas, and (2) each grid area can be considered to be covered by only one type of vegetation. The first such model was suggested by Dickinson et al. (1984). This model included descriptions of interception of rainfall, changes of moisture storage in a two-layer soil, root water uptake, stomatal resistance, canopy radiation, boundary-layer resistance of the canopy and, aerodynamic resistance of the soil and canopy. It has come to be known as the Biosphere–Atmosphere Transfer Scheme (BATS).

Sellers et al. (1986) reviewed developments in this type of modelling and proposed a "simple biosphere model" (SiB). Both these models allow the different types of landscapes, soils, and biophysical and geometrical characteristics of vegetation to be taken into account, but this leads to the inclusion of an enormous quantity of parameters, which must either be measured or calibrated. For example, the Sellers' canopy model contains 49 tunable constants. The necessity for and applicability of such complex models of evapotranspiration for GCMs, has been questioned (McNaughton and Jarvis, 1983). Indeed both BATS and SiB are constructed mainly on the basis of *a priori* assumptions which have not been verified using data observed at a point or at the small plot scale.

Numerical simulations with "big leaf" models in GCMs have shown that this parameterisation of evapotranspiration can improve, at least qualitatively, the simulation of most meteorological fields. For example, comparisons of simulations made using the "bucket" model and the SiB model in the spectral GCM of the National Meteorological Centre of the USA for 50 day summer periods and 30 day winter periods, have shown that the SiB model allows (1) evapotranspiration over territories with vegetation and (2) the decrease of precipitation over continents, to be estimated more accurately (Sato et al., 1989).

In the work of Henderson-Sellers (1990), more convincing results were described. The BATS model was included in the GCM of the National Centre for Atmospheric

Research of USA (horizontal resolutions of 4.5° latitude by 7.5° longitude) for the simulation of the hydrological cycle over Australia for a period of 3 years. Land characteristics (porosity, hydraulic conductivity of saturated soil, root distribution, heat capacity, albedo) were assigned for 18 different types of landscape, 8 classes of soil texture and 4 colours of land. Good agreement between simulated and observed fields of monthly air temperature, precipitation, evapotranspiration and vertical heat fluxes was observed. However, simulations of monthly runoff fields appeared to be very poor. Unsatisfactory runoff simulations based on the use of "big leaf" models in GCMs are described in a number of other publications (e.g. Thomas, 1990). It seems likely that these results are not atypical and can be caused by a relatively crude method of assigning values to land constants (using the principle of "physical realism") which does not take into account small and meso-scale variations in land characteristics and which can play the leading role in runoff formation. It must be stressed that runoff formation evolves in time much more rapidly than evapotranspiration and the absence of a procedure for determining effective precipitation (i.e. that which generates runoff) can give, for many regions, large errors in calculating evapotranspiration.

Close agreement with observations of large-scale meteorological fields, can be achieved in GCM simulations by proper tuning of some model parameters (e.g. maximum soil moisture capacity), but, as a result, a significant distortion of other important characteristics of climate systems (for instance, river runoff) can occur. The application of sophisticated physically-based models of hydrological cycle components (such as BATS or SiB) can simplify the choice and calibration of a few key parameters, but does not guarantee good simulations of the climate system as a whole.

It seems that similar results can be obtained, if detailed physically-based models of runoff formation are used but, because of process averaging over large areas, the values of average or effective parameters will differ significantly from their measured values. One such approach has been discussed by Monin (1979), who proposed dividing the continents using the boundaries of river basins, within which water movement was described as two-dimensional overland flow to main river channels. The river basin area was considered as a large slope. In this case ground-water runoff is also described as two-dimensional movement along the basin slope. The rainfall excess was determined as the difference between the rainfall and infiltration rates; the latter was calculated using the equation of vertical soil moisture movement including capillary and non-capillary filtration. Such a representation means that, inside the model, almost all variables and coefficients are "effective" values which can be tens of times more or less than their real values. The physical meaning of the output variables of the model can be preserved.

The fitting of relevant effective parameters can guarantee the representation of the necessary effects in GCMs in spite of poor descriptions of hydrological processes at small-scale levels. But the representation within such a model of spatial change in land characteristics and the effective significance of model parameters, will approximate accurately their real significances, only if (a) the spatial resolution of the model increases, (b) the new information on land characteristics etc becomes available, and (c) the description of processes improves.

The potential for improving the description of the hydrological cycle in present-day GCMs depends to a marked degree on whether general hydrological relationships can be found for large areas. Large-scale geographical generalisations, which allow relationships to be established between water balance components and the factors affected by them, have been available for a long time. These attempts were most successful for large time intervals and first of all, for multi-year mean values. Among such generalisations are maps of annual and seasonal runoff, of runoff ratios, different types of regression dependencies relating runoff with precipitation, temperature and soil moisture indexes, height and slope of a river basin, and relative areas of forests, lakes and swamps. The accuracies of such dependencies are as a rule not high, but sometimes they can be increased by including in the empirical dependencies, physically-based *a priori* assumptions or values. For instance, in many cases, the dependence between snowmelt runoff and quantities such as snow water equivalent and indices of autumn soil moisture and depth of frozen soil, can be stable enough for large areas. For some periods the relationships between soil moisture and air temperature or vapour pressure deficit are sufficiently strong. Many formulae for calculating potential evapotranspiration for months, seasons and multi-year periods are well known.

One of the first attempts at building a coupled water balance and river water transport model at the continental scale was made by Vörösmarty et al. (1989). The model was constructed as part of a description of global biogeochemistry, which determined the choice of both time–space scales and the continent: South America, where Amazonia was the region of special interest for investigating the influence of rainforest and large tropical rivers on global change.

South America was represented by more than 5,700 0.5° (latitude/longitude) grid cells, where, for each grid cell, data on long-term mean monthly temperature, precipitation and potential evapotranspiration, as well as characteristics of vegetation, soils and topography, were available. This information has been transformed into predictions of soil moisture, evaporation and runoff hydrographs. Mean monthly temperatures were obtained from archived data. Precipitation and potential evapotranspiration data were taken from the UNESCO Atlas of World Water Balance (Korzoun et al., 1977). In this atlas, potential evapotranspiration has been estimated from air temperature, radiation balance and soil moisture content. For the calculation of actual evapotranspiration, the techniques proposed by Thornthwaite and Mather (1957) were used. The field capacity, which is necessary for this technique, was determined as a function of soil texture and vegetation; six types of texture and three types of vegetation with different root depths for different types of texture were used. As a result 11 different values of field capacity and their spatial distributions were generated.

The monthly values of runoff determined by the water balance method were used for hydrograph routing in the Amazon/Tocantins river system. The simulated Amazon Basin was comprised of 1936 grid cells, and the Tocantins River, 315 cells. For each cell, a predominant direction of flow was determined to form the network topology of the river system. Channel flow was represented by linear reservoir models. Each grid cell was considered as a reservoir with the same characteristic residence time and it was lengthened or shortened depending on the geometry of influent and effluent rivers. To predict floodplain inundation, it was assumed that an exchange between channel and floodplain took place whenever the monthly discharge exceeded a specified fraction of long-term mean annual flow. The characteristic residence time and two constants taking into account floodplain inundation, were fitted using the calculated and observed long-term mean annual hydrographs at 6 selected sites. An analysis of the results from the water balance models and a comparison of calculated and observed hydrographs at intermediate sites showed that acceptable agreement had been achieved in spite of a number of relatively crude assumptions. For example, the influence on runoff of ground water and of change in soil moisture was not been taken into account; too simplified a model of runoff routing was used, etc. More recently, the same models have been applied successfully to water balance and water transport simulation for another large tropical river, the Zambezi in southern Africa (Vörösmarty and Moore, 1991).

However, the accuracy and the predictive ability of these models cannot be high for many regions because of the short-term persistence characterising a number of hydrological processes. For instance, runoff generated by rainfall for most European rivers mainly depends on antecedent precipitation conditions for a period of less than two weeks. The short-term rise of air temperature can cause intensive snowmelt and a significant change in the runoff hydrograph. In many cases, the temporal distributions of precipitation and temperature during monthly or seasonal time intervals play a more important role in the formation of runoff and evaporation than the magnitudes of average values of these variables.

One of the possible ways of parameterising the hydrological cycle in GCMs is to couple them with lumped conceptual models of runoff formation, after improving the descriptions of areal evapotranspiration in these models. But conceptual models usually include a lot of empirical parameters, which are determined by calibration based on the observed output data of hydrological systems. As the size of basin area and its heterogeneity increases, the range of variations of such empirical parameters can also increase.

One of the ways of improving the estimation of the response of a hydrological system, which is considered to be lumped, is to introduce instead of lumped parameters, their statistical distributions; these distributions can be interpreted as proportions of basin area with different values of the parameters. Such distributions can be determined from empirical data or from *a priori* assumptions, and, in hydrology, there is considerable experience of using this approach. Nemec and Becker (1987) were apparently the first to suggest applying such statistical distributions to describe hydrological processes in GCMs. Entekhabi and Eagleson (1989) have shown that, if it can be assumed that statistical gamma distributions of rainfall and soil moisture apply inside the grid area of a GCM, simple analytic expressions for runoff and evapotranspiration can be derived. The main question which arises when using this approach, concerns the "stability" of the applied distributions, where stability in the context of GCMs relates to the possibility of using them for several grid areas. It seems that the spatial distribution of soil moisture (type of distribution and its parameters) strongly depends on antecedent meteorological conditions and can be quite different after dry and wet periods. It is clear that the stability of such statistical distributions is associated with the large-scale variance of the values under consideration (large-scale homogeneity). It seems that the size of present-day GCM sub-grid areas is too large to permit the use of statistical distributions based on proportions of area for most important hydrological constants.

The most natural way of developing GCM parameterisations of the hydrological cycle for a horizontal scale of 100–300 km is by aggregating the results of models with a finer resolution. But to do this, the problem of disaggregating input data for sub-grid areas needs to be solved, which, in many cases, is no less simple than the construction of the hydrological model itself.

4. Modelling of the terrestrial hydrological cycle at a horizontal scale of 30–100 km

Current hydrological practice in describing the hydrological cycle for catchments with areas from hundreds to thousands of square kilometres relies primarily on lumped conceptual models of runoff formation. However, in describing hydrological cycle within GCMs, the potential of conceptual models is limited, because of the shortage for many river basins of the information needed for the calibration of large numbers of empirical parameters. Relationships for determining parameter values for such models based on measured topography and river system characteristics are usually found to have low accuracy and reliability. Certain problems are also caused by different physical interpretations of variables and constants used in GCMs and conceptual models. Furthermore, the various types of remotely-sensed data that are becoming more widely available can be used more effectively for determining the physical constants of hydrological systems, than for estimating the aggregated parameters of conceptual models.

All these reasons, and the experience accumulated over the last decade in the development and application of physically-based hydrological models, suggest that these models can be a dominant basis for the construction of models of the hydrological cycle for areas of continental scale. It seems that there are two main tasks to be undertaken in adapting these models for larger spatial scales and coupling with GCMs:

(1) The most well developed physically based hydrological models, such as the Système Hydrologique Européen (SHE) (Abbott et al., 1986 a,b; Bathurst, 1986) or the models of the USSR Water Problems Institute (Kuchment et al., 1983, 1986, 1989) were created in the first instance to describe runoff formation and so they do not incorporate sufficiently detailed descriptions of vertical soil–vegetation–atmosphere transfers (SVAT) for coupling with GCMs.

(2) It is necessary to pay more attention to sub-grid-scale processes associated with small scale changes in land characteristics and hydrometeorlogical values, inside the discretisation domain used for the numerical integration of the partial differential equations or the representation of different land types.

The development and validation of physically based models have been mainly carried out for small river basins with good observational data; the largest basin to which such a model has been applied is the river Sosna with a catchment area of 16,000 km^2. It is necessary to find such sizes of the grid spacing, for which it is possible (a) to represent large-scale changes in topographic, soil and vegetation characteristics, (b) to guarantee the stability of relationships used for sub-grid effects inside areas considered as lumped, and (c) to keep computing time within reasonable limits.

If the size of grid cells of GCMs is taken as 50 km $\times$ 50 km, this means that the characteristic time of horizontal intergrid movement of the main mass of water (travel time of river runoff) is approximately 2–2.5 days. If the river channel system is divided into 15–20 reaches (this is usually sufficient) then the time interval for input data (local inflow) must be approximately 3 hours or less. For the calculation of the subsurface response and ground-water runoff, the time interval can be 12–24 hours or more. It seems that the calculation of ground-water runoff and ground-water levels can be carried out for larger areas and time periods than is the case in the calculation of surface flow. In calculating average infiltration and evapotranspiration, it is desirable to use the shortest time period of the GCM model (15–20 minutes).

Taking into account the different time scales of overland, subsurface and ground-water flows and the results of research on the sensitivity of these to a number of relief, soil and vegetation constants, it is possible to suggest the following approach to describing the hydrological cycle for space scales of up to 10,000 km^2.

4.1. Basin schematisation

For the descriptions of overland and subsurface flow, the basin is divided into slopes along which one-dimensional flow to the channel system network can occur. On these slopes quadrangular sections with soil and topography characteristics, which are uniform as far as possible, are chosen. These sections serve as finite elements for numerical integration of the overland and subsurface flow equations.

The basin channel system which interacts with ground water is represented on a two-dimensional rectangular grid. But the dendritic structure of the river system must be taken into account and relevant corrections to the lengths of river segments introduced to preserve the general length of the river system. The same two-dimensional grid is used for ground-water modelling.

4.2. Input to finite elements

The point precipitation intensity P_Λ is considered to be exponentially distributed over the fraction *e* of the finite element area:

$$\phi_1(P_\Lambda) = \frac{e}{E(P)} \exp\left[-\frac{e\,P_\Lambda}{E(P)}\right]$$

where $E(P)$ is the finite element average intensity and $\phi_1(P_\Lambda)$ is the area density function of P_Λ.

The value e can be determined from a relation between the intensity and area of rainfall.

4.3. Soil moisture transfer and point evapotranspiration

Vertical moisture in unsaturated soil can be described by the differential equation:

$$\frac{\partial q}{\partial \tau} = \frac{\partial}{\partial \zeta}\left[\Delta(q)\frac{\partial q}{\partial \zeta} - K(q)\right] - \Sigma(q,\zeta)$$

where τ is time, ζ is depth and q is the volumetric moisture content; $\Delta(q)$ is the soil diffusivity; $K(q, K_\sigma, q_{max})$ is the soil hydraulic conductivity; $\Sigma(q,\zeta)$ is the rate of water extraction by roots; K_σ is the hydraulic conductivity of saturated soil; and q_{max} is the porosity. $\Delta(q)$ can be expressed by empirical formulae which depend on K_σ, q_{max} and the maximum hydroscopicity j_{mh}. $\Sigma(q,\zeta)$ depends on the root length λ_r and the minimum stomatal resistance P_{sm}. A more detailed description of this model can be found in Kuchment and Startseva (1991).

4.4. Rainfall excess

The average rainfall excess for each finite element is defined as:

$$E(\theta) \int_0^1\!\int_0^1\!\int_0^1\!\int_0^1\!\int_0^1\!\int_0^1 \left[P_\Lambda(E(P)) - E_a(K_\sigma, \rho_{sm}, \lambda_r) - \Gamma_m - I(K_\sigma, j_{mh}, q_{max})\right]$$

$$\times\, \phi_1(P_\Lambda)\, \phi_2(K_\sigma)\, \phi_3(\rho_{sm})\, \phi_4(\lambda_r)\, \phi_5(j_{mh})\, \phi_6(q_{max})\; d\phi_1\, d\phi_2 \,\ldots\, d\phi_6$$

where E_a is the point evapotranspiration; ρ_{sm}, the stomatal resistance; I, the rate of infiltration; ϕ_2, ϕ_3, ..., ϕ_6 are the area distribution functions for K_s, ρ_{sm}, λ_r, j_{mh} and q_{max}. Γ_m is canopy interception.

4.5. Average surface depression interception for each finite element

$$E(\theta_s) = E(\theta)(1-\rho) - \Pi_{max}\left[1 - \exp\left(-\frac{\eta_s}{\Pi_{max}}\right)\right]$$

$E(\theta_s)$ is the input to surface overland runoff; η_s is the depth of overland flow; ρ is the fraction of closed depressions dependent on η_s; and Π_{max} is the maximum depression losses.

4.6. Surface overland flow

$$\frac{\partial \eta_s}{\partial \tau} + \frac{\partial E\Theta_s}{\partial \zeta} = E(\theta_s) + E_H(\theta_\iota)$$

$$\Theta_s = \nu_s^{-1} \iota_0^{1/2} \eta_s^{5/3}$$

for area fraction ϕ_7 for which $(P_\Lambda) > I + E_a - \Gamma_m$ or $\eta_\iota > H_\iota$; η_ι is the subsurface flow depth; H_ι is the maximum η_ι; ι_0 is the slope; and ν_s is the Manning roughness.

For each finite element

$$E(\Theta_s) = \int_0^1 \Theta_s \phi_8(\iota_0)\, d\iota_0$$

where Θ_s is the point overland discharge; $E_H(\theta_\iota)$ is the input from subsurface flow when $\eta_\iota > H_\iota$; $\phi_8(\iota_0)$ is the distribution of areas with different values.

Subsurface flow in the case of deep and shallow ground-water tables and flow-routing of surface water can be treated using established models. A model of winter hydrological processes and of snowmelt runoff formation are described in Kuchment et al. (1983, 1986).

5. Opportunities for developing parameterisations of the hydrological cycle for horizontal resolutions of less than 30 km

Detailed observations of water balances components for catchments with areas from tens to hundreds of square kilometres have shown that the spatial variabilities of runoff and evapotranspiration are very large, even for small plots. Apart from heterogeneities in relief, soils, and vegetation, an important role for small basins is played by the prevalence of some types of runoff formation mechanisms or certain specific conditions, which have an influence on vertical moisture fluxes. Carefully designed process experiments at the small plot scale are required for better understanding of such mechanisms.

The choice of the optimum horizontal resolution is of fundamental significance. From the point of view of the influence of hydrological processes and air-moisture transfer, the level of detail adopted in representing the spatial variability of vertical moisture fluxes depends on the ranges of horizontal mixing in the surface turbulent boundary layer. If it is assumed that the depth of this layer is about 100 m and that the horizontal mixing homogenises the transfer of air properties on horizontal scales of the order of ten times more than this depth, it is sufficient to take into account only heterogeneities on scales of 1 km or larger. It is also known that 10 km appears to be the minimum scale, beyond which inhomogeneities in land surface properties can trigger specific meso-scale atmospheric circulation systems, which have a definite impact on mean large-scale transfers between the land surface and the atmosphere. However, these estimates are probably too crude and they can vary significantly for different geographical regions.

The most difficult task in coupling GCMs with fine resolution hydrological models is the spatial disaggregation of meteorological data inputs. For small scales this disaggregation must take into account such factors as the exposition of river slopes, the spatial variation of hydrophysical soil constants and albedo, snow distribution, pre-history of moisture and heat regimes of the soil, etc. To assign all these characteristics, it is necessary to change essentially the methodology of measurement of the meteorological variables which can be used for calibration of meso- or micro-meteorological models. It seems that the difficulties in disaggregating meteorological values calculated in GCMs and used in models of the hydrological cycle can be the main limitation in choosing a minimum horizontal resolution for the description of hydrological processes.

The transitions in spatial scales which are necessary to take into account the influence of various types of vegetation and soil on vertical moisture and heat exchanges, requires more sophisticated research on the peculiarities of evapotranspiration from different ecological systems; up to now, most attention has been paid to agricultural systems.

The problem of parameterising sub-grid effects for small scales also appears. This demands research into micro-scale changes of land surface characteristics. Indeed the construction of models of the terrestrial global hydrological cycle with a horizontal resolution of 1–10 km seems to be impossible without satellite or aircraft remote sensing systems. These systems must provide both direct measurements of land-surface constants and data for determining model parameters.

Some of these problems are traditional key issues of scientific hydrology, but most of them require joint cooperation with specialists in different fields of geophysics, biophysics and space technology. The Global Energy and Water Cycle Experiment (GEWEX), being conducted as part of the World Climate Research Programme, provides an excellent opportunity for such cooperation. HAPEX, FIFE, KUREX and other large-scale field experiments have been carried under the umbrella of GEWEX and have given many useful results, which can improve our understanding of the processes of vertical water and energy transfer. But too little has been done overall to get new experimental information for the construction of macro-scale hydrological models.

Acknowledgements

This work was performed during the author's collaboration with Professor P.E. O'Connell at the University of Newcastle-upon-Tyne in the Spring of 1991.

References

Abbott, M.B. et al., 1986a. An introduction to the European Hydrological System Système Hydrologique Européen (SHE), 1: History and philosophy of a physically-based distributed modelling system. J. Hydrol., 87: 45–59.

Abbott, M.B. et al., 1986b. An introduction to the European System-Système Hydrologique Européen (SHE), 2: Structure of a physically-based distributed modelling system. J. Hydrol., 87: 60–79.

Bathurst, J.C., 1986. Physically-based distributed modelling of an upland catchment using the Système Hydrologique Européen. J. Hydrol., 87: 80–102.

Budyko, M.J., 1948. Evaporation under natural conditions. Gimiz, Leningrad (JPST, Jerusalem, 1969).

Dickinson, R.E., 1984. Modelling evapotranspiration for three-dimensional global climate models. In: Climate processes and climate sensitivity. Am. Geophys. Union, Geophys. Monogr., 29, vol. 15.

Entekhabi, D. and Eagleson, P.S., 1989. Land surface hydrology parametrisation for atmospheric general circulation models including subgrid scale spatial variability. J. Climate, 2: 816–831.

Giorgi, F. and Mearns, L.O., 1990. Simulation of regional climate change: a review. Rev. Geophys., 29: 2.

Henderson-Sellers, A., 1990. Evaluation for the continent of Australia of the simulation of the surface climate using the Biosphere–Atmosphere Transfer Scheme (BATS) coupled into a global climate model. Climate Res., 1: 43–62.

Korzoun, V.J. et al., 1977. Atlas of World Water Balance. UNESCO, Paris.

Kuchment, L.S. and Startseva, Z.P., 1991. Modelling moisture transfer in the soil–canopy–atmosphere systems. IAHS Publ. No. 204.

Kuchment, L.S., Demidov, V.N. and Motovilov, Ju.G., 1983. Formation of river runoff. Publ. "Nauka", Moscow (in Russian).

Kuchment, L.S., Demidov, V.N. and Motovilov, Ju.G., 1986. A physically-based model of the formation of snowmelt and rainfall-runoff. IAHS Publ. No. 155, pp. 27–37.

Kuchment, L.S. et al., 1989. Modelling moisture transfer in systems "soil–vegetation–atmosphere" for hydrological tasks. Water Resour., 2: 32–49 (in Russian).

Mahlman, J.D. and Umscheid, L.J., 1987. Comprehensive modelling of the middle atmosphere. Influence of horizontal resolution. In: C. Visconti and R. Cardia (Editors), Transport Processes in the Middle Atmosphere. D. Reidel, Dordrecht.

McNaughton, K.G. and Jarvis, P.G., 1983. Water deficit and plant growth. Advances in Ecology, Vol. VIII. Academic Press, New York, N.Y., pp. 1–47.

Monin, A.S., 1979. Hydrodynamic equations of terrestrial hydrology. Rep. USSR Acad. Sci., 246 (5): 1063–1066 (in Russian).

Nemec, J. and Becker, A., 1987. Macroscale hydrologic models in support to climate research. IAHS Publ. No. 168.

Sato, N. et al., 1989. Effects of implementing the simple biosphere model in General Circulation model. J. Atmos. Sci., 46: 2757–2782.

Sellers, R.I., Mintz, I., Sud, Y.C. and Dalcher, A., 1986. A simple biosphere (SiB) model for use within general circulation models. J. Atmos. Sci., 43: 505–531.

Thomas, G., 1990. The regional hydrologic impacts of global climate change: the role of climate models. Palaeogeogr., Palaeoclimatol., Palaeoecol., 82: 343–368.

Thornthwaite, C.W. and Mather, J.R., 1957. Instructions and tables for computing potential evapotranspiration and the water balance. D. Drexel Inst. Technol. Publ. Clim., X(3).

Vörösmarty, Ch.J. and Moore, B., 1991. Modelling basin-scale hydrology in support of physical climate and global biogeochemical studies: an example using the Zambezi river. Surv. Geophys., 12: 271–311.

Vörösmarty, Ch.J. et al., 1989. Continental scale models of water balance and fluvial transport: an application to South America. Global Biogeochem. Cycles, 3 (3): 241–265.

Chapter 9

A rainfall–runoff scheme for use in the Hamburg climate model

L. DÜMENIL and E. TODINI

Abstract

With a view to increasing the realism of the description of land surface processes in the Hamburg climate model, a scheme is introduced that accounts for the partition of rainfall at the ground between infiltration into the soil and surface runoff. The scheme was adapted and calibrated for the catchment of the river Arno and takes the heterogeneity of the land surface within a grid area into account. For the use in the general circulation model it has been expanded to account for stronger runoff components in regions of mountainous terrain. The GCM simulated annual globally averaged runoff is found to be in good agreement with observations.

1. Introduction

Many scientific institutions worldwide are involved in studying global climate and global change. The general public also is becoming increasingly aware of these issues. Any scientific statement concerning these problems must be founded on sound scientific grounds, and must take into account state-of-the-art knowledge on all the important processes of the climate system. In this field, experimentation can only be carried out with the help of powerful coupled models of the general circulation of the atmosphere and oceans (OAGCMs). Until recently only relatively low horizontal and vertical resolution could be afforded in climate modelling. Apart from this problem, the areas which require research in the atmospheric part of these models are i.a. the representation of clouds and the various components of the hydrological cycle. In particular, the hydrological processes at the land surface, have so far only been represented in OAGCMs in a rather rudimentary way.

As an initial step to expand the level of complexity of land surface processes in the Hamburg OAGCM (Fischer, 1987), we shall introduce a scheme that accounts for the partition of rainfall between infiltration into the soil and surface runoff. The so-called Arno (or Xinan-Jiang) scheme was derived from catchment considerations and takes the heterogeneity of the land surface within a grid area into account. It is designed to compute, for a given rainfall event, (a) the amount of water at the surface that will infiltrate into the soil in the regions where the soil is not yet saturated, and (b) the amount of water that will run off in the regions where the soil has reached saturation.

2. Runoff in the context of general circulation models

2.1. Motivation

When the new second generation Hamburg model ECHAM was developed, the need for improvement of this particular aspect of the model was identified. This was motivated by two facts. The first is due to the task of coupling atmosphere and ocean models. Because of the presence of the ocean, several processes need to be considered in the coupled atmosphere–ocean system that are traditionally neglected in weather forecasting models.

In contrast to weather forecasting models, sea-ice cover is required as a prognostic variable, as is the freshwater flux. The latter is a variable that influences both the thermohaline circulation in the ocean and the formation of sea-ice via its influence on salinity. It is therefore a critical variable for the evolution of climate. Over the oceans, the freshwater flux is normally taken as $P - E$, precipitation minus evaporation. Over land the process is more complex. The water reaching the surface during precipitation events is distributed locally between infiltration into the soil and immediate runoff at the surface. The water is then transferred to the rivers and eventually to the sea. Depending on the local soil and orographic as well as other conditions, drainage from the wet soils may also add to the river flow, and some water percolates from the soil to the water table underneath.

The first generation of the Hamburg climate model (Fischer, 1987) consisted of a low-resolution version of the European Centre for Medium Range Weather Forecasts' (ECMWF) weather prediction model. Due to their specific task and their relatively short integration times of not more than 10 days, the results of weather prediction models are influenced by land surface processes to a lesser degree than climate models which are typically integrated for decades. Land surface processes are therefore only computed in a rather rudimentary way. In this model, the time evolution equation of soil water was updated according to the scheme of Deardorff (1972). This scheme is not acceptable for climate studies because it relies on an updating process based on present-day climatology and does not take hydrological information into account. The amount of water that goes into runoff is normally not of interest in a ten-day forecast.

The Deardorff scheme assumes the same type of force-restore method for water in the soil as for heat conduction. Evaporation loss and input due to precipitation events are components of the soil water budget. The full equations are approximated by two prognostic equations for two layers of soil water and allow a diffusive exchange between these layers. A value specified from climatology (Mintz and Serafini, 1981) is used as the lower boundary condition. If the prescribed soil field capacity for either layer is reached, runoff may occur. This gives the old scheme the character of the "bucket scheme" that was implemented e.g. in the GFDL OAGCM by Manabe (1969). For climate research in particular, the aspect of a fixed climate boundary condition at all times is unacceptable.

While the processes associated with soil hydrology are relatively well understood and may be modelled successfully and in great detail at the point, element and small catchment scale, they cannot yet be modelled at the same level of detail at the scale of very large river catchments (continental scale). OAGCMs on the other hand will still be limited, for a considerable length of time, to relatively large grid sizes, since their application, in the context of climate change, requires typical integration times of the order of 100 years. Although the resolution may well improve over the next decade, enabling us to resolve at least medium-sized river catchments, a process of up-scaling of our knowledge of hydrologic processes will still have to be initiated in order to incorporate these processes in OAGCMs at sufficient detail for global climate forecasts.

In the meantime, relatively simple but physically based approaches have to fill this gap in the representation of the hydrological cycle. It is required that they can take the heterogeneity of the land surface within a grid area into account.

The motivation to replace the Deardorff scheme was therefore two-fold: In the climate model, all processes have to be formulated such that they are independent of present-day climatology, because this would otherwise preclude the simulation of e.g. paleo-climates. Second, the concept of the Deardorff scheme may well be appropriate for heat diffusion and in medium-range forecasts, but it is not based on the knowledge of hydrological processes and their time-scales. In order to incorporate more hydrology into their parameterisation scheme, an approach different from the simple "bucket" model was adopted in the UK Meteorological Office model (Warrilow et al., 1986). Here, the Richards' equation is simplified and applied at each grid-cell. For us, this approach did not seem to be a good alternative, one reason being the fact that this type of equation is only valid at a single point. Specific, and so far undeveloped, techniques are required in order to extend this scheme beyond the microscale.

Although more sophisticated rainfall–runoff schemes are available in hydrological engineering applications [Stanford Watershed Model IV (Crawford and Linsley, 1966), National Weather Service Soil Water Accounting Model (Burnash et al., 1973), SSAR (Rocwood and Nelson, 1966), Continuous API (Sittner et al., 1969), TANK (Sugawara et al., 1976), SHE (Abbott et al., 1986)], their implementation in OAGCMs is strongly inhibited by the lack of data for verification and initialisation of such schemes on the global scale. Even though many applications in hydrology provide excellent data coverage at specific points, more data will be needed on the distributions of soil characteristics, the heterogeneity (and how it influences the mean over a grid area) of the land surface, the distribution of water with depth in the soil and the movement of the water table with the seasons. All of these are at present poorly understood, but field campaigns are being initiated in order to improve the required data base. For the time being, it is therefore recommended to keep the scheme as simple as possible and the number of tuning parameters at a minimum.

The catchment scheme to be introduced below is a simple extension of the "bucket" scheme, from the point of view of an atmosphere modeller, and a simplification of the more sophisticated hydrology schemes that are employed in water resources research and engineering. The scheme has been successfully applied to the catchment of the river Arno in Italy. Input and output and the basic assumptions are such that the model is independent of the present-day climate and should be equally applicable at all model resolutions (i.e. grid cells of 5.0° × 5.0° or 1.0° × 1.0°).

2.2. The ECHAM Model

As mentioned above, the atmospheric part of the Hamburg climate model has evolved in several steps. The ECMWF model was used initially (Fischer, 1987). It was replaced by the ECHAM model, where most of the original parameterisations for cloud cover, radiation, horizontal diffusion and heat flux in the soil and the water reservoir, were replaced by revised schemes. In particular the treatment of the boundary conditions over land was modified in order to provide a better physical basis for climate, i.e. long-term, integrations. The spectral formulation of the original model was, however, retained. A detailed description of this so-called ECHAM version of the model and its climatology is available in Roeckner et al. (1989, 1992).

All model results, which will be discussed later, are taken from the ECHAM model integrated at T21 resolution (equivalent to a grid spacing of 5.625° longitude, i.e. typical grid size of 600 km × 600 km). In the ECHAM model, the following land surface processes are represented. Precipitation may reach the ground as rain or snow. At the surface, a single reservoir for soil water evolves with time using infiltration and evaporation as input. The Arno scheme has been integrated interactively in the GCM to specify how much rainfall is available for runoff and infiltration, respectively. The model calculates evaporation from the atmospheric demand which is then modified by a simple vegetation scheme (Blondin, 1989) to give the actual evapotranspiration. This is subtracted from the soil water reservoir. In cold conditions precipitation may fall as snow and will accumulate on the ground. When this snow-pack begins to melt, the melt water goes directly into surface-runoff if the ground underneath is still frozen. If the ground is thawing, the snow melt water also enters the Arno scheme.

After the computation of infiltration, drainage and runoff at the grid cell, the water is subjected to a linear advection scheme (Sausen et al., 1991) which collects the runoff in large river catchments and transfers it to the oceans.

3. Description of the Arno scheme

Over a large catchment (or a grid cell), saturated and unsaturated regions are unevenly distributed. There may be areas in the vicinity of rivers and streams that are saturated most of the time, whereas slopes and areas with certain soil conditions never reach saturation at all. If it rains, the regions with small field capacities will produce runoff at once, whereas in regions where field capacities are relatively large, rain water arriving at the surface may infiltrate for a certain period of time, i.e. until the local soil field capacity is filled and saturation is reached. Only from then onwards, will further rainfall produce runoff. Soils that have reached soil water values close to saturation, experience a depletion of soil water by the processes of drainage and percolation to the water table. Runoff and drainage are collected and transported horizontally in catchments of various sizes and eventually take the precipitated water to the oceans where it enters as fresh water inflow.

Bucket schemes, as they have been used so far, operate on the assumption that the soil water available in a catchment area is evenly distributed over the whole area. This means that no runoff will occur in a catchment until the whole area reaches saturation. Usually bucket schemes do not account for drainage and percolation.

The scheme presented here tries to describe the process of infiltration and runoff (including drainage) based on simple hydrological experience. The main point is that, unlike the bucket model, it takes the heterogeneous distribution of soil water capacity into account.

3.1. Derivation of the equations

The basic foundation of the model is derived from the integration over the whole catchment of the equation of water balance at a point on the ground.

The original model was developed by R.J. Zhao at the East China Technical University of Water Resources at Nanking (now HeHai University), and is known as the Xinan-Jiang model (Zhao, 1977).

Subsequently, the original scheme was modified by Todini (1988) in order to take into account, more effectively, the effect of drainage on soil depletion. Drainage, together with evapotranspiration, is responsible for the highly non-linear behaviour of the catchment response.

The equation that expresses the local mass balance at one point in the basin is:

$$r = \begin{cases} p - (w - w_0) & \text{if } p > w - w_0 \\ 0 & \text{if } p \leq w - w_0 \end{cases} \tag{3.1}$$

where r is the local surface runoff (mm), p is the local precipitation (mm), w is the local storage capacity (mm), and w_0 is the initial uniform water content in the soil (mm) within the basin.

The total volume of runoff produced over the entire basin may be expressed by the average depth R, i.e. by the integral:

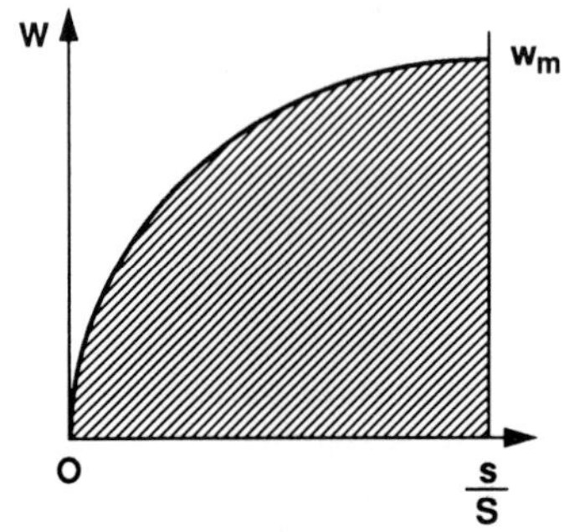

Fig. 1. Schematic of the distribution of soil water capacities at the point scale. The shape of the curve is determined by the parameter b and the total grid cell soil water content is W_{m}.

$$R = \frac{1}{S}\int_0^S r\,\mathrm{d}s = \frac{1}{S}\int_0^S [(p - (w - w_0)]\,\mathrm{d}s \tag{3.2}$$

where R is the mean integral value of the runoff, and S is the total area of the basin.

The total soil capacity of the catchment consists of the aggregate of many different local soil water capacities that depend on the type of soil. In other words, for a given catchment, the storage capacity is not represented by a unique value as is assumed in a bucket model, but by a set of values with a probability density distribution $f(w)$.

It is possible to define the "storage capacity distribution curve" as a curve which defines the percentage s/S of the basin area in which the storage capacity is less than or equal to an assigned value w. It is expressed as follows:

$$\frac{s}{S} = 1 - \left(1 - \frac{w}{w_{\mathrm{m}}}\right)^b = F(w) = \int_0^w f(\xi)\,\mathrm{d}\xi \tag{3.3}$$

where w is the storage capacity value, $0 < w \leq w_{\mathrm{m}}$; w_{m} the maximum soil water capacity in the basin; b the characteristic parameter of the basin; and $F(w)$ the cumulative probability function of the random variable w.

The shaded area between the curve and the x-axis in Figure 1 represents the average storage capacity of the entire basin, which is indicated by the symbol W_{m} as:

$$W_{\mathrm{m}} = \int_0^1 w\,\mathrm{d}\left(\frac{s}{S}\right) = \frac{1}{S}\int_0^S w\,\mathrm{d}s \tag{3.4}$$

Figure 2 illustrates how the rainfall is distributed between runoff and infiltration. Let us assume that initially the soil water reservoir does not contain any water. Precipitation event 1 makes a certain amount of water available which is P_1. As we have no measure as yet on how the precipitation is distributed over the catchment area, we assume an even distribution, indicated by the straight horizontal line

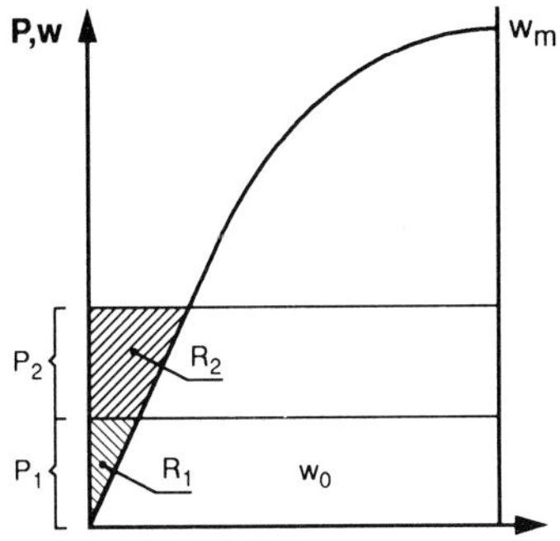

Fig. 2. Distribution of rainfall between runoff and infiltration for two precipitation events with the same amount of water. $P_{1,2}$ are evenly distributed over the grid cell, P_1 begins at zero initial water content, P_2 at initial water content W_0.

in Figure 2. The point where this line intersects the curve for w, separates the saturated and unsaturated parts of the catchment. Where P_1 exceeds w, runoff may occur, where $P_1 < w$, the water infiltrates into the ground. The total water content of the reservoir in the ground integrated over the catchment is now W_0.

The next precipitation event P_2 starts from the condition of a partly filled reservoir. If we assume $P_2 = P_1$, the area of saturation where runoff may occur is now much larger and consequently a larger percentage of P_2 appears as runoff than was the case for P_1.

The calculation of runoff may then be expressed in the form:

$$R = P - (W_m - W_0) \tag{3.5a}$$

for:

$$P > (1 + b)W_m\left(1 - \frac{W_0}{W_m}\right)^{\frac{1}{b+1}}$$

or by:

$$R = P - (W_m - W_0) + W_m\left[\left(1 - \frac{W_0}{W_m}\right)^{\frac{1}{b+1}} - \left(\frac{P}{(b+1)W_m}\right)^{b+1}\right] \tag{3.5b}$$

for:

$$P \leq (1 + b)W_m\left(1 - \frac{W_0}{W_m}\right)^{\frac{1}{b+1}}$$

where the upper case quantities indicate integral mean values.

These equations of state must be associated with the balance equation by which the mean water content in the soil is updated.

The equation takes the form:

$$W_0(t + dt) = W_0(t) + P(t) - E(t) - R(t) - D(t) - I(t) \tag{3.6}$$

where $E(t)$ is the loss through evapotranspiration between t and $t + dt$, $D(t)$ is the

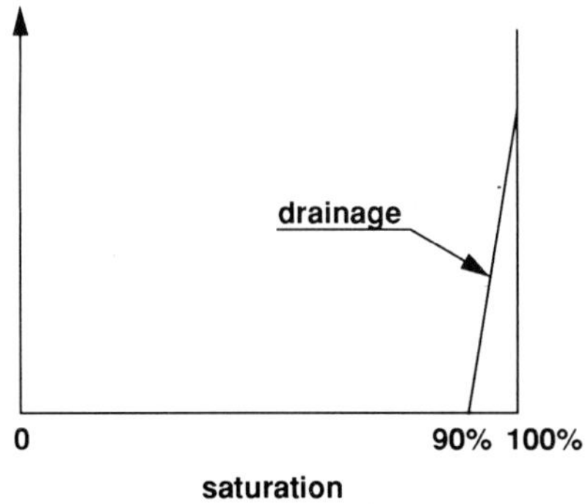

Fig. 3. Drainage of water in the soil adds to the local runoff. Drainage occurs as a linear function of total soil water in the range of 90–100% saturation.

loss through drainage between t and $t + \mathrm{d}t$, and $I(t)$ is the percolation loss to deep water tables between t and $t + \mathrm{d}t$.

The drainage term $D(t)$ is generally linked to the water content of the soil (Figure 3) in non-linear form and can be expressed as follows:

$$D(t) = \begin{cases} \alpha D_{\max} \dfrac{W_0}{W_\mathrm{m}} & W_0 \leq W_\mathrm{D} \\ \alpha D_{\max} \dfrac{W_0}{W_\mathrm{m}} + (1-\alpha) D_{\max} \left(\dfrac{W_0 - W_\mathrm{D}}{W_\mathrm{m} - W_D} \right)^c & W_0 > W_\mathrm{D} \end{cases} \tag{3.7}$$

where α is the percentage of maximum drainage at maximum water content W_m; $D_{\max}$ the maximum drainage; $\mathrm{W_D}$ the water content threshold value; and c the exponent of the variation law: $c = 1$ linear, $c = 2$ quadratic. Upper case quantities indicate mean integral values.

The quantity $I(t)$ varies only slightly over time when compared to the other terms; nevertheless a non-linear behaviour may be assumed for this, which may be expressed as follows:

$$I(t) = \begin{cases} 0 & W_0 \leq W_\mathrm{i} \\ \beta(W_0 - W_\mathrm{i}) & W_0 > W_\mathrm{i} \end{cases} \tag{3.8}$$

where W_i represents the water content threshold value for calculating deep infiltration.

Finally, the term $E(t)$ may be derived from the evapotranspiration ET_0 reference value by means of suitable crop coefficients and taking into account the actual quantity of water present in the soil.

The total outflow $R_\mathrm{T}(t)$ produced by the precipitation $P(t)$ is finally:

$$R_\mathrm{T}(t) = R(t) + D(t) + B(t) \tag{3.9}$$

where $B(t)$ is the outflow originating from deep water tables.

3.2. Modifications for use in the OAGCM

As was shown in the previous section, the scheme can be reduced to a set of equations which only uses information on the average quantity of initial water content and current water content of the total soil reservoir. It is therefore a parameterisation in the true sense, because it expresses sub-grid scale variability and processes in terms of the averaged large-scale value. This is a familiar concept which is used for many physical processes that cannot be resolved explicitly in general circulation models.

The scheme also fulfils another requirement of OAGCMs because of its simplicity: the same set of calculations is executed at each land point in the OAGCM. This is an important pre-requisite for an efficient use of today's vector processing computers.

The scheme was originally designed with specific catchments in mind (Xinan-Jiang and Arno river; Todini, 1988). The equations are, however, general and simple enough to allow the application of the scheme to a grid cell in the general circulation model. There is also no general limitation to the size of the area to which the scheme may be applied.

A further advantage of the scheme is that it has a very small number of parameters which need to be specified. In a specific catchment the range of these parameters may be established from observational data to which the model is applied and which can be used for model calibration. Figure 4 shows part of the catchment of the Arno river in northern Italy to which the scheme was originally adapted for river-flow forecasting. The gauge data from this river was used to calibrate the model (Figure 5a). Calibrated in this way, the model produces a forecast which is in very good agreement with observed gauge data taken from another period (Figure 5b; Todini, 1988).

For application in the OAGCM we have adopted the following guidelines. In the OAGCM all land surfaces have at present the same thermal characteristics everywhere, if they are not covered by snow or ice. We have therefore assumed the same value of W_m for the maximum grid-cell soil-water saturation-capacity at all OAGCM land points.

In a catchment, a typical average value for the parameter b would be $b = 0.2$. We chose to modify this parameter at grid cells with steep orography. In this way, the scheme is expected to produce more than average runoff in mountainous regions. This orography correction to b is made dependent on the model resolution:

$$b = \max\left[\frac{\sigma_h - \sigma_0}{\sigma_h + \sigma_{max}} : 0.01\right] \tag{3.11}$$

Values of b lie within the range 0.01 to 0.5, when the minimum and maximum standard deviations of orography, σ_0 and σ_{max}, are used, namely:

$$\sigma_0 = 100 \text{ m} \tag{3.12}$$

$$\sigma_{max} = 1500 \text{ m at T21}, \quad \text{or } \sigma_{max} = 1000 \text{ m at T42 resolution}$$

(a)

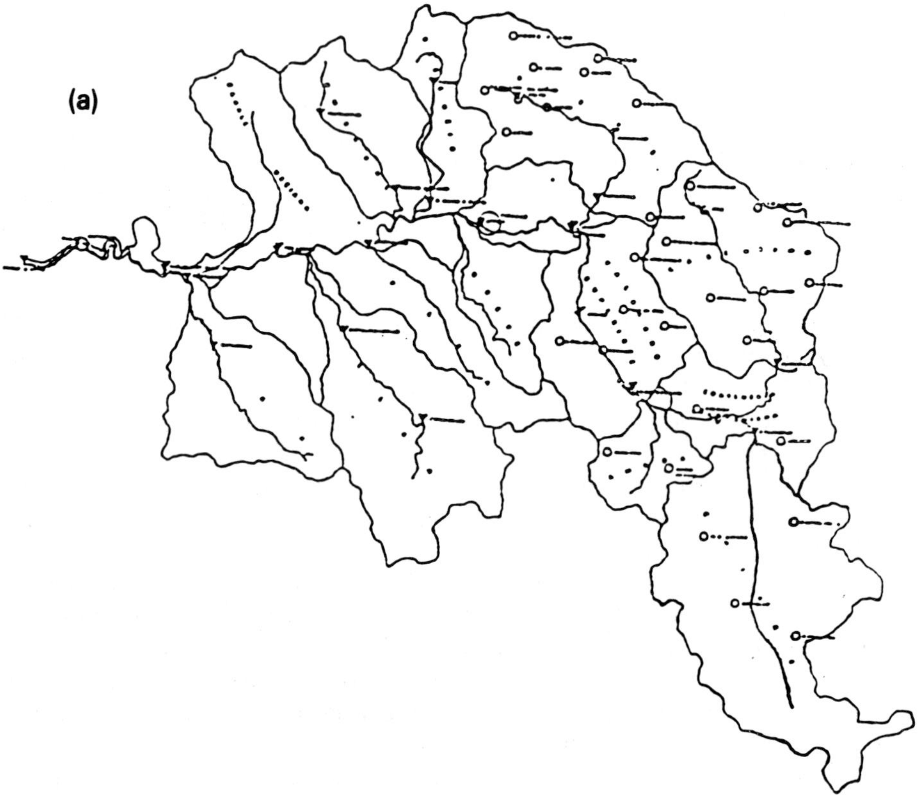

Fig. 4. a. The catchment of the Arno river in northern Italy and the hydrological gauge network.

Furthermore, the scheme is extended to OAGCM areas covered with snow or ice. From the equation for the OAGCM snow pack on the ground, we can deduce the amount of water that becomes available during the melting of the snow. If the soil underneath the snow pack is frozen, all melt water goes directly to runoff. If the soil is not frozen, the melt water is subjected to the runoff/infiltration scheme as described in section 3.1.

The drainage component of the runoff at a grid cell R_D is computed from equation (3.7) if the soil has reached more than 90% of its field capacity and if the soil is not frozen:

$$R_D = \begin{cases} d_{\min} \dfrac{W_s}{w_{\max}} & \text{if } W_s < W_{DR} \\ d_{\min} \dfrac{W_s}{w_{\max}} + (d_{\max} - d_{\min}) \left(\dfrac{W_s - W_{DR}}{W_{\max} - W_{DR}} \right) D & \text{if } W_s \geq W_{DR} \end{cases} \tag{3.13}$$

with $d_{\min} = 0.0005$ mm/hr, $d_{\max} = 0.05$ mm/hr, $W_{DR} = 0.90 W_{\max}$ and $D = 1.5$.

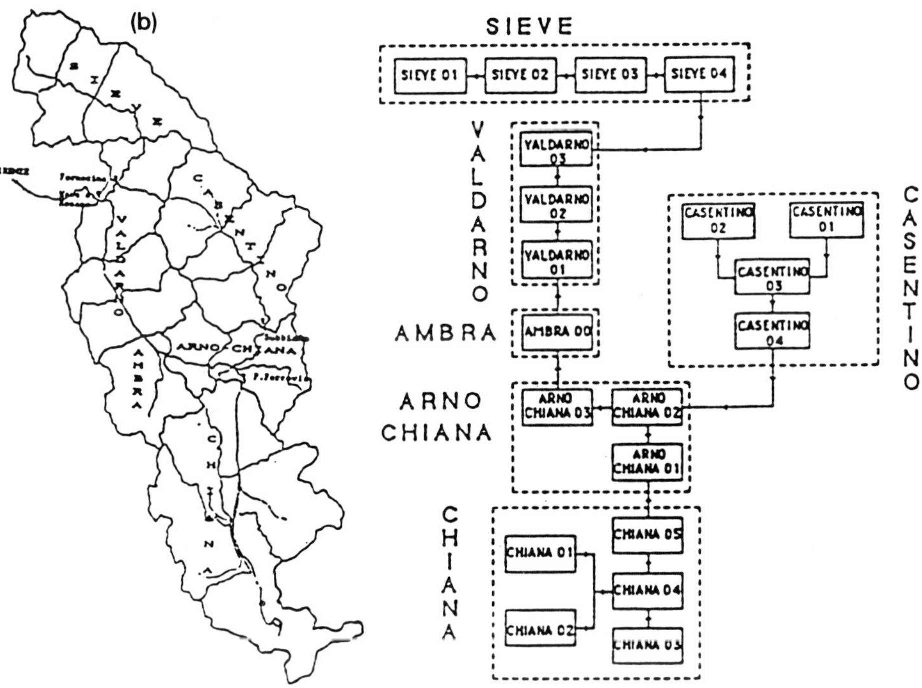

Fig. 4 (continued). b. Schematic of the sub-catchments used in flood forecasting with the Arno model.

The percolation component, I, that is contained in the catchment version of the scheme is neglected in the present version of the scheme, because no extra information on the water table is available in the model.

4. The simulated hydrological cycle

The model parameters and processes associated with the hydrological cycle are an important part of any OAGCM simulation. Due to the lack and insufficiency of climatological data for these fields on the global scale, the model generated fields are, however, difficult to validate in detail. In the following, we shall qualitatively assess the precipitation field, the surface runoff generated by the Arno scheme, the soil water reservoir, and the output from the river runoff scheme during the seasonal cycle. In this first attempt to verify model results, we shall try mainly to establish the plausibility and consistency of the simulated fields which are all inter-related.

All results are monthly means taken from a 20-year integration using the ECHAM2-T21 model as described in section 2. In the simulation, diurnal and seasonal cycles of solar insolation are applied.

A first crude comparison of the parameters of the hydrological cycle with climatology is given in Figure 6 (Roeckner et al, 1992). Evaporation, precipitation

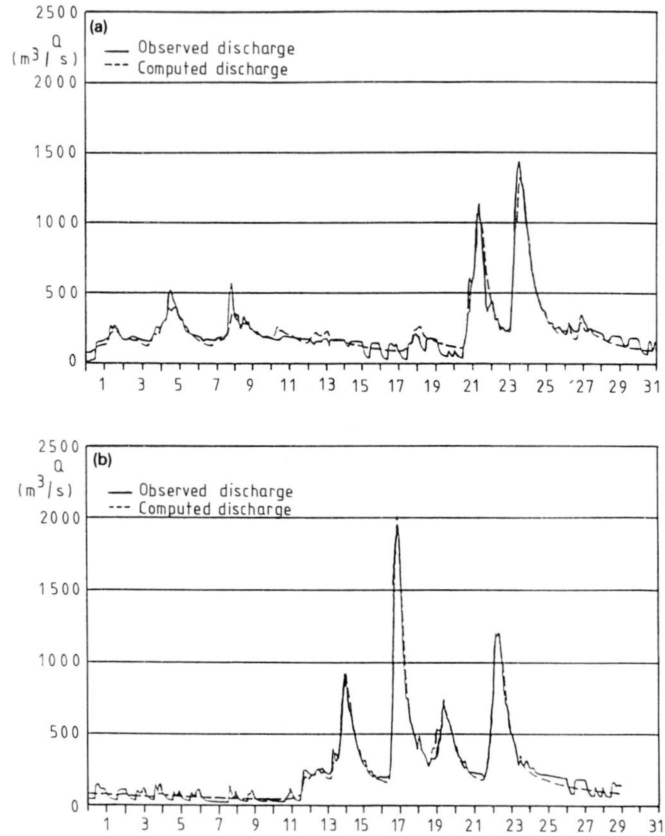

Fig. 5. Application of the Arno scheme to hydrological data. (a) Calibration of the model using data from December 1959. (b) Hind-casting of the flooding in February 1960: model results and verification data.

and runoff are compared to observations by Baumgartner and Reichel (1975). The globally averaged long-term annual mean values of the hydrological parameters simulated over land are in good agreement with the climatology, but are smaller than observed. Compared to the climatology, there is less precipitation arriving at the continental surface and less evaporation. Although the amount of freshwater that is made available to the oceans by the continental runoff is close to the observed, a relatively large amount of precipitation is taken out of the cycle and stored in the permanent snow packs on the continents such as Greenland and Antarctica. The model does not account for the freshwater input into the ocean that occurs directly from glaciers by the process of calving.

4.1. Precipitation

Figure 7 displays the model simulated field of total precipitation for January and July. Model results are compared to the climatology by Jaeger (1976) and

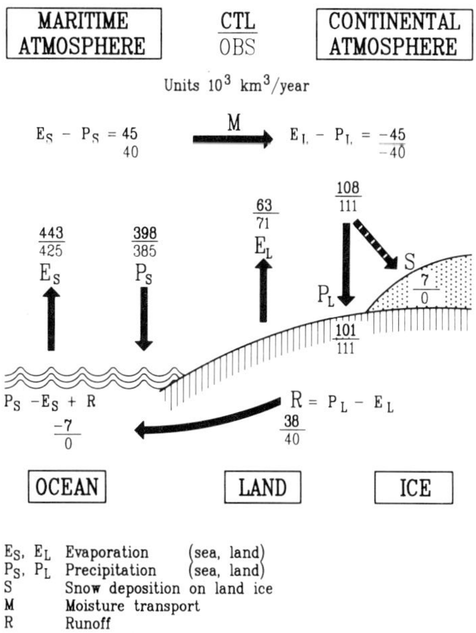

Fig. 6. Global annually averaged hydrological cycle from a 20-year model simulation (ECHAM2, upper values), and the climatology by Baumgartner and Reichel (1975, lower values).

a more recent one by Legates (1987) and Legates and Willmott (1989). A good representation of the global field of precipitation by the model's dynamics and convection parameterisation schemes is, of course, the pre-requisite for a successful simulation of the runoff process. The precipitation field simulated by the ECHAM2-T21 model compares favourably with the Jaeger climatology and with results from other models of the same resolution, but lacks many details. (The structure of this field is smoothed due to Jaeger's data sampling at 5° intervals which is about the same resolution as that of the present model.)

In general, the model has a tendency to broaden precipitation events of convective or orographic origin so that some of the maxima that are present in the climatology are not simulated. In January, e.g., there is no maximum at the southernmost tip of South America, and in Africa, the ITCZ is not well pronounced so that the dry regions extend almost to the equator. In South America the model produces much less rainfall than is observed due to systematic model errors which also occur in other low resolution models. Where precipitation is dominated by large-scale processes, e.g. in the Northern Hemisphere storm tracks, the precipitation either as rain or snow is steered by the cyclone paths. In Europe in winter, cyclones penetrate into the continent in a realistic way and a much better snow pack is simulated than in North America (Behr and Dümenil, 1991). In particular the behaviour of cyclone tracks in North America in the spring is affected by systematic model errors so that the precipitation in the middle of the continent is overestimated.

(a)

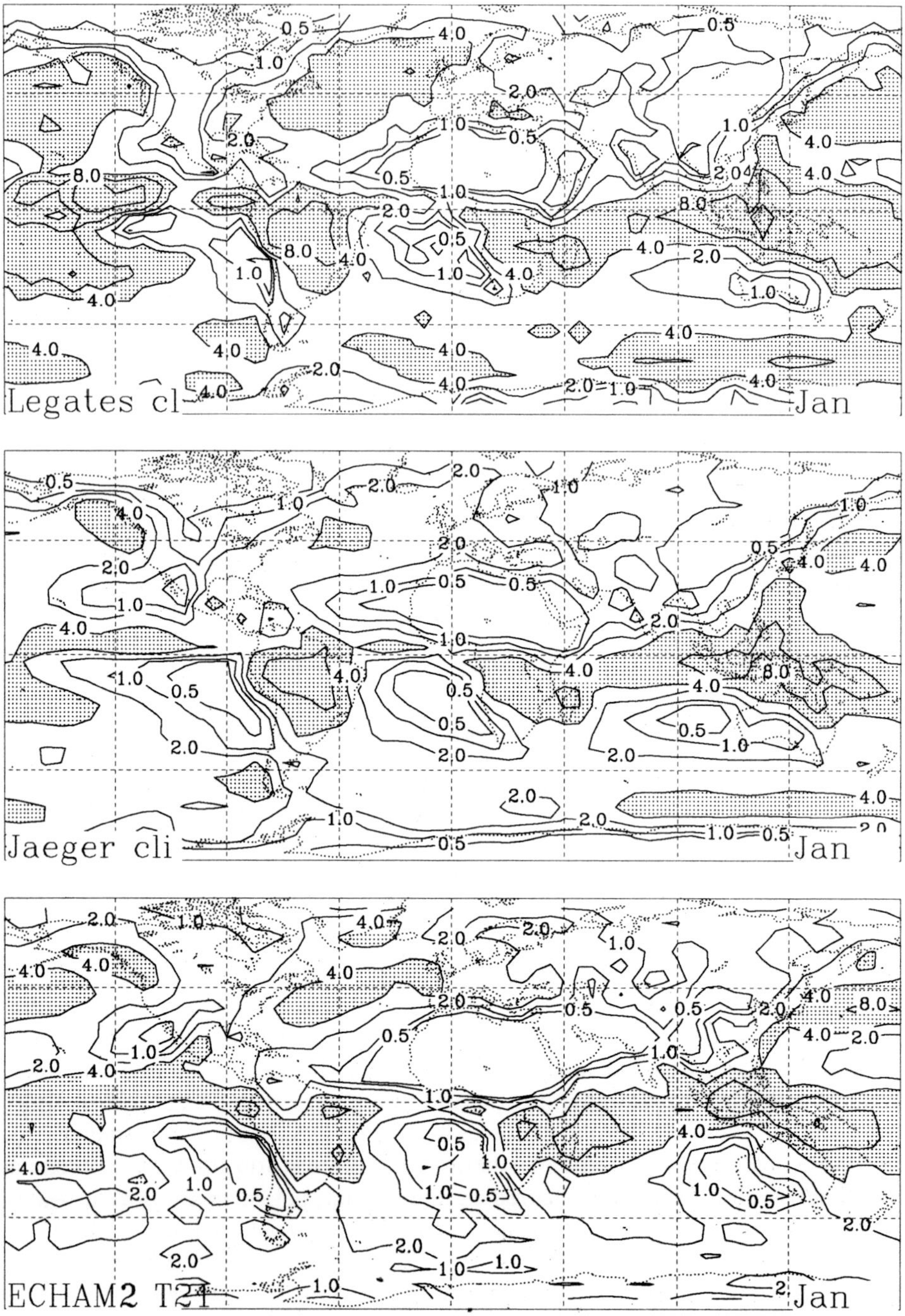

Fig. 7. a. Horizontal distribution of total precipitation for the month of January. Climatologies by Legates (1987) and Legates and Willmot (1989; top), Jaeger (1976; middle) compared to monthly mean values from a 20-year integration with the ECHAM model (bottom).

(b)

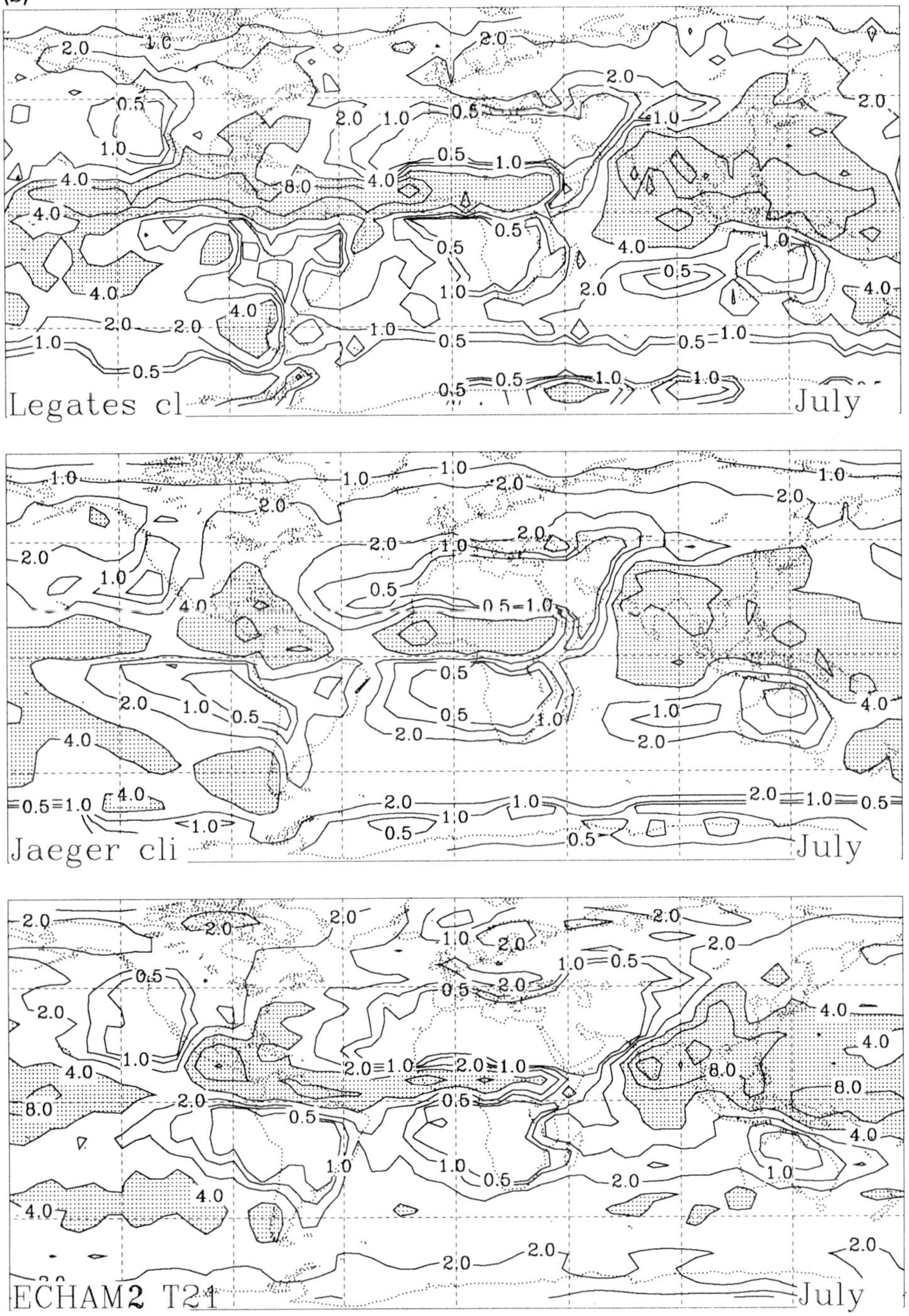

Fig. 7 (continued). b. As Figure 7a but for July.

In July, the larger-scale sinking motions that cause dryness over the subtropical continents, are well represented by the model in the Southern Hemisphere, but too far extended in the Northern Hemisphere. The ITCZ, although much more pronounced than in winter, does not extend far enough to the north. In some aspects the model ITCZ compares very well with Legates' (1987; Legates and Willmott, 1989) climatology. Also in the monsoon region the model simulated precipitation is in better agreement with this latter dataset (not shown).

4.2. Local runoff

The quantity of local runoff (Figure 8) at the surface is derived from the precipitation field according to the Arno scheme as described in section 3. The rainfall part of the precipitation is subjected to the scheme immediately and also drainage is computed at each time step. The amount of precipitation that falls as snow during the cold season is first stored in the snow pack until the melting season begins and is therefore subjected to the scheme with considerable delay.

We are at present not in a position to verify the time evolution of the local surface runoff for the continents in the OAGCM by direct comparison with observations. There is no information on how much runoff will occur at a certain grid cell for a given amount of precipitation. Even if this type of information were available, the Arno scheme would not stand up to a local verification, because of the current restriction on the variation of the parameter b as a function of orography only. So far no variation of land surface (soils) characteristics is applied. Considering this, we can only expect the simulated runoff field to represent — in a qualitative sense — the rainfall/runoff coefficients that mainly reflect mountainous regions and a consistent picture of the more trivial process of larger runoff in areas with higher amounts of rainfall and/or snow melt and soil water. Figure 8 shows that the scheme represents this very well, without an undue overestimation of the orography component.

During the annual cycle (Figure 8) we see that the model represents maximum runoff in the tropical regions, into which the Intertropical Convergence Zone migrates. In winter, Europe and the coastal areas of the North American continent, experience large runoff due to the high amounts of rainfall. Where the precipitation falls as snow, e.g. in Siberia, almost no runoff occurs during the winter months. It is not until April that large parts of the Northern Hemisphere continents experience high surface runoff as a result of the melting of the snow pack. The behaviour of the model snow pack was diagnosed to be in better agreement with observations in Eurasia than in North America (Behr and Dümenil, 1991). The snow cover there is deeper than observed and therefore more water is available for snow melt in the spring.

The change in the season from spring to summer is very marked in the tropics and the northern parts of the American and Eurasian continents. The ITCZ has now moved north of the equator, and on the Indian subcontinent and in South

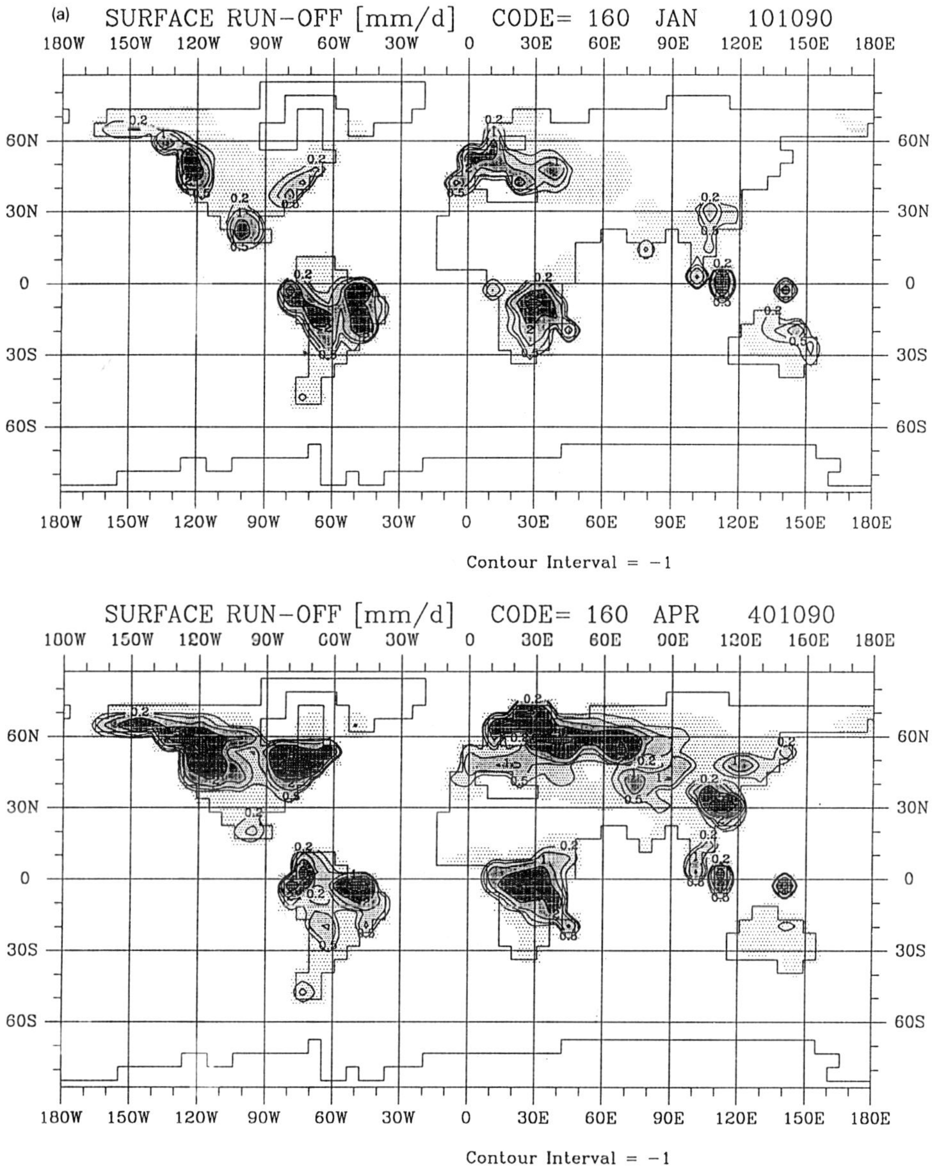

Fig. 8. a. Local surface runoff computed by the Arno model in the OAGCM. Values are from a 20-year integration with the ECHAM2 model for January (top) and April (bottom).

East Asia the monsoon has set in. Overall, the seasonal cycle of model generated local runoff seems to be acceptable. The global mean value of local runoff over the continents at 38×10^3 km/year is in good agreement with observations by Baumgartner and Reichel (1975).

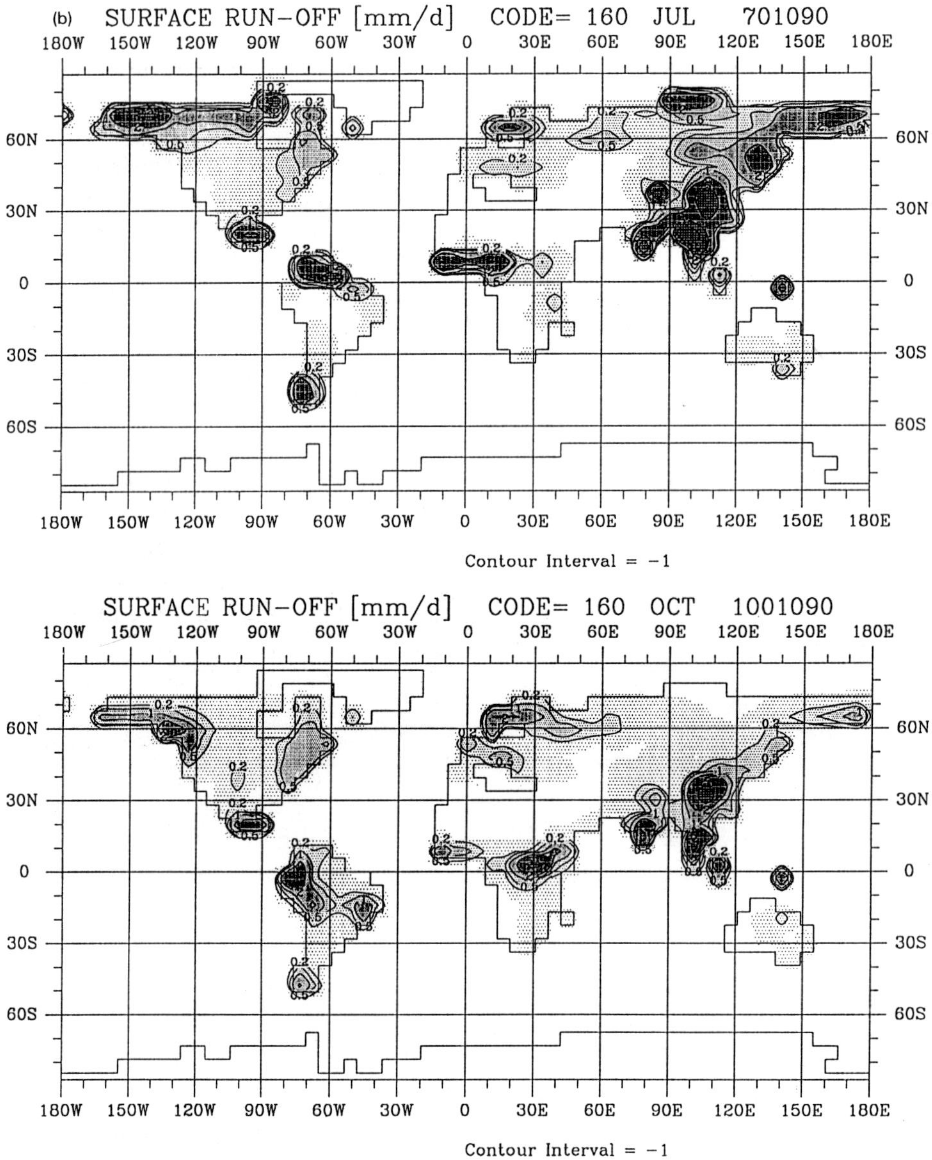

Fig. 8 (continued). b. Values for July (top) and October (bottom).

4.3. Soil water

The surface runoff is also steered indirectly by the amount of evaporation that occurs in a grid cell, because evaporation is a component of the soil water budget. An evaporation climatology has also been derived by Mintz and Serafini (1981) but it is not independent of the soil water climatology by the same authors. We shall therefore discuss only the latter. The performance of the soil water reservoir cannot yet be validated directly against observations on the OAGCM scale. Mintz and Serafini (1981) have derived a global field of large-scale "soil water" from observed precipitation and evaporation derived from a simple model. Figure 9 compares the results from the OAGCM with this surrogate climatology. In addition we now have to realize that systematic errors may be introduced by the atmospheric demand for water vapour (general circulation) and the formulation of the parameterisation scheme for evapotranspiration.

With a few exceptions, the model simulation of the general features of the seasonal cycle of soil wetness is in good agreement with the climatology. In the climatology we distinguish between the dry desert regions in the subtropics caused by descending air motion and the wet tropical areas where precipitation is abundant. In the model simulated field, the very dry regions are also well positioned in the subtropics (e.g. in July), but are less dry than in the climatology.

The largest differences with reference to the soil water reservoir occur in those regions, where the model simulates a melting snow pack. There are two reasons for this disagreement: first, the model snow pack is deeper than observed and the melting is delayed (as e.g. in North America), and, second, the climatology does not account for the process of snow melt. It was derived using precipitation data and an evapotranspiration model only (Serafini, pers. commun., 1990).

During the winter months, the South American continent is much drier in the model than in the climatology. This is due to a systematic model error in the precipitation field caused by the general circulation and is not sensitive to changes in the parameterisation of soil hydrology.

Soil wetness is also an OAGCM parameter that cannot be observed directly. For the purpose of OAGCM validation, surrogates are needed. If the local runoff in the model can be collected in large river catchments by a river routing or river runoff scheme (e.g. Sausen et al., 1991) the time delay between local runoff at a grid cell and river gauge measurements can be represented. We can then compare model results with river gauge data measured at the mouths of the largest rivers as an additional verification tool.

4.4. River gauge data

Validation of monthly mean values against river gauge data can provide a useful independent measure of model performance. It may complement the data from climatologies which are, at present, insufficient and unreliable. This type of data is being collected in a global data base and an effort is being made to create a comprehensive climatological dataset for the purpose of OAGCM validation for the twenty largest rivers in the world (Global Runoff Data Centre, Koblenz, F.R.G.). Until an agreed dataset (corrected for irrigation, sufficiently long time series, etc.) becomes available, validation studies can only be considered as preliminary attempts.

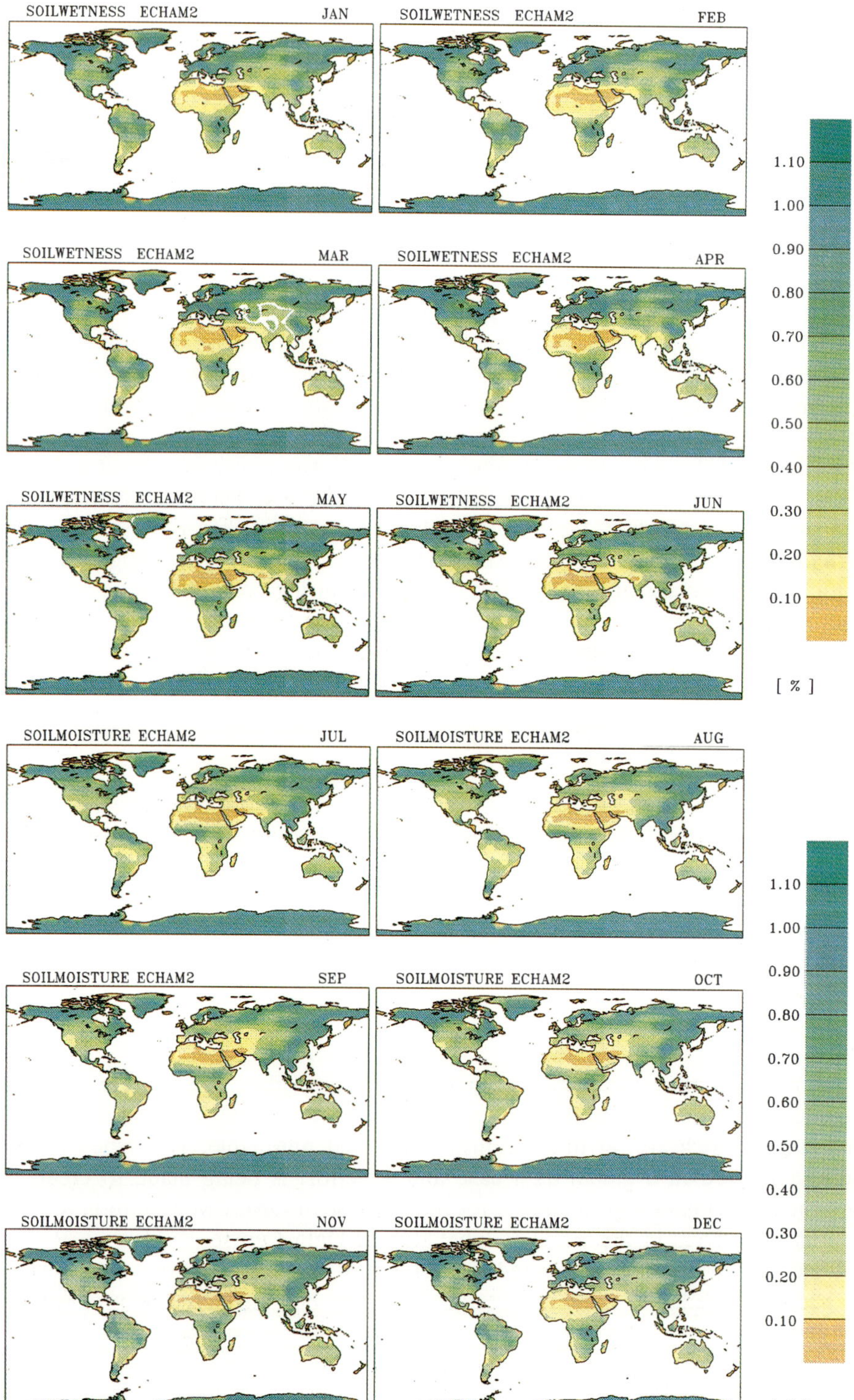

Fig. 9. Soil wetness as computed by the ECHAM2 model during the annual cycle.

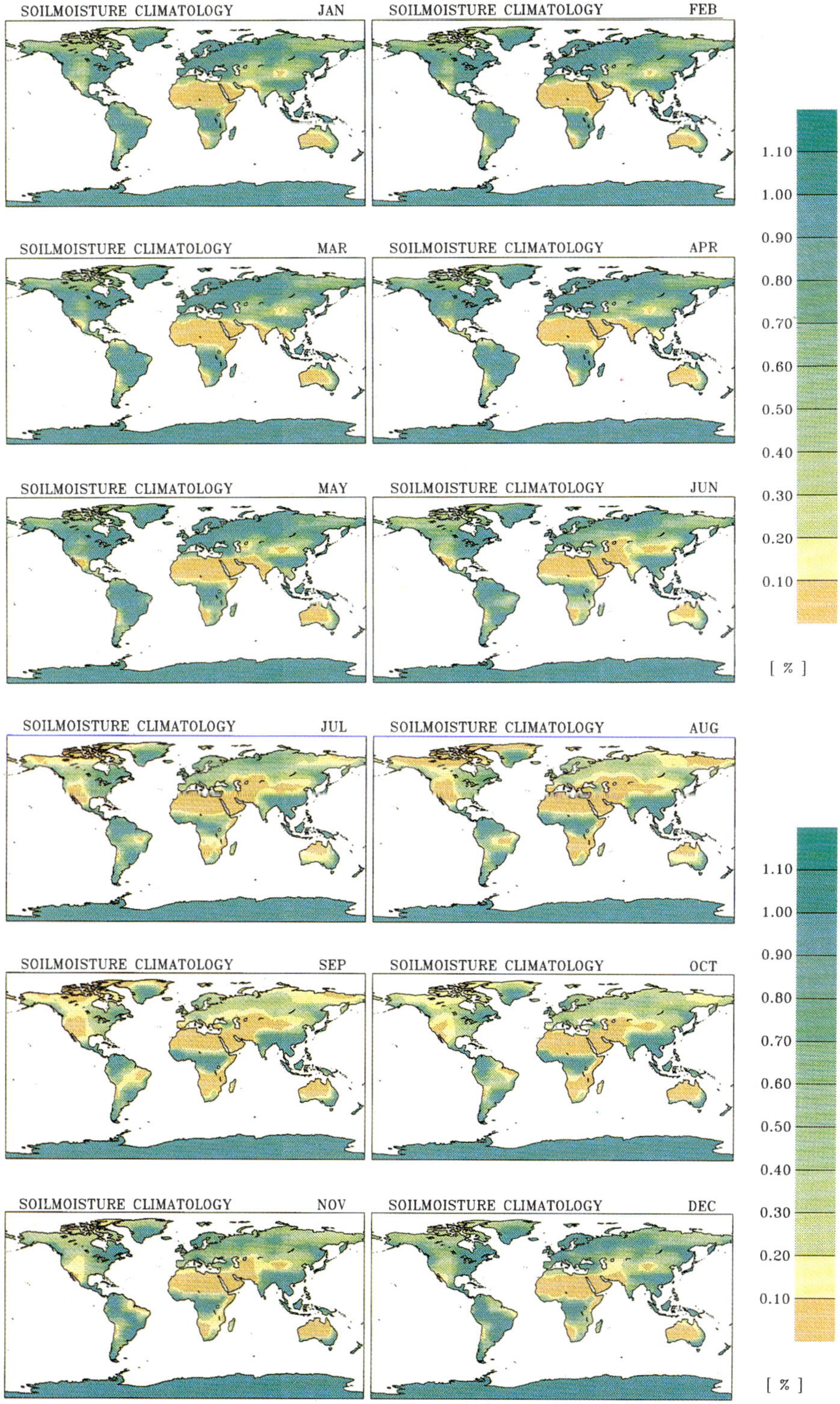

Fig. 9 (continued). Climatology is redrawn from Mintz and Serafini (1981).

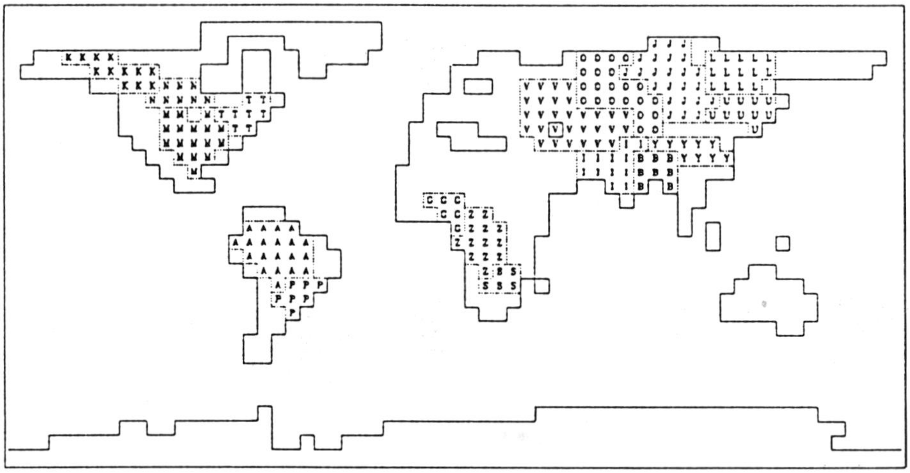

Fig. 10. Catchments of the largest rivers which are consistent with the ECHAM2-T21 model and its orography from Sausen et al. (1991).

Figure 10 shows the catchments which are consistent with the ECHAM2 model configuration. They may not always agree with the real catchments. Figure 11 shows the simulated and observed annual cycles of water, reaching the mouths of several large rivers. Considering only this small selection of rivers with large catchments we find that the model results vary between very good agreement (Lena, Jenissei) and total disagreement (Niger, St. Lawrence). There is a considerable variation in the observed maxima which have only been averaged in a crude way for this study and are not corrected for irrigation, etc.

Most of the deficiencies are, however, explained by the systematic model errors in the precipitation field. For example, the large amount of water transported by the Amazon river is also well represented by the model, but during the second half of the annual cycle, the model produces less rainfall than observed, so that the runoff is reduced almost to zero.

The fact that the Asian Monsoon season is well represented by the ECHAM2 model is reflected in the good agreement found for the Indus river and the Brahmaputra/Ganges.

The strong disagreement seen for the Niger river in Africa will require further investigation. It is likely that irrigation is the cause of the bias in the observed data available to us. The main part of the difference is, however, due to the bad representation of rainfall in the model's ITCZ in winter.

In general, the rivers dominated by snow melt in the spring are in good agreement with observations in Eurasia. We have already remarked that the model's deeper snow pack in North America disappears later in the year, than observed from other independent sources using snow data (Behr and Dümenil, 1991). Hence the disagreement with the river gauge data in the spring in this case is not surprising. In Eurasia where the simulated snow values agree much better with observations,

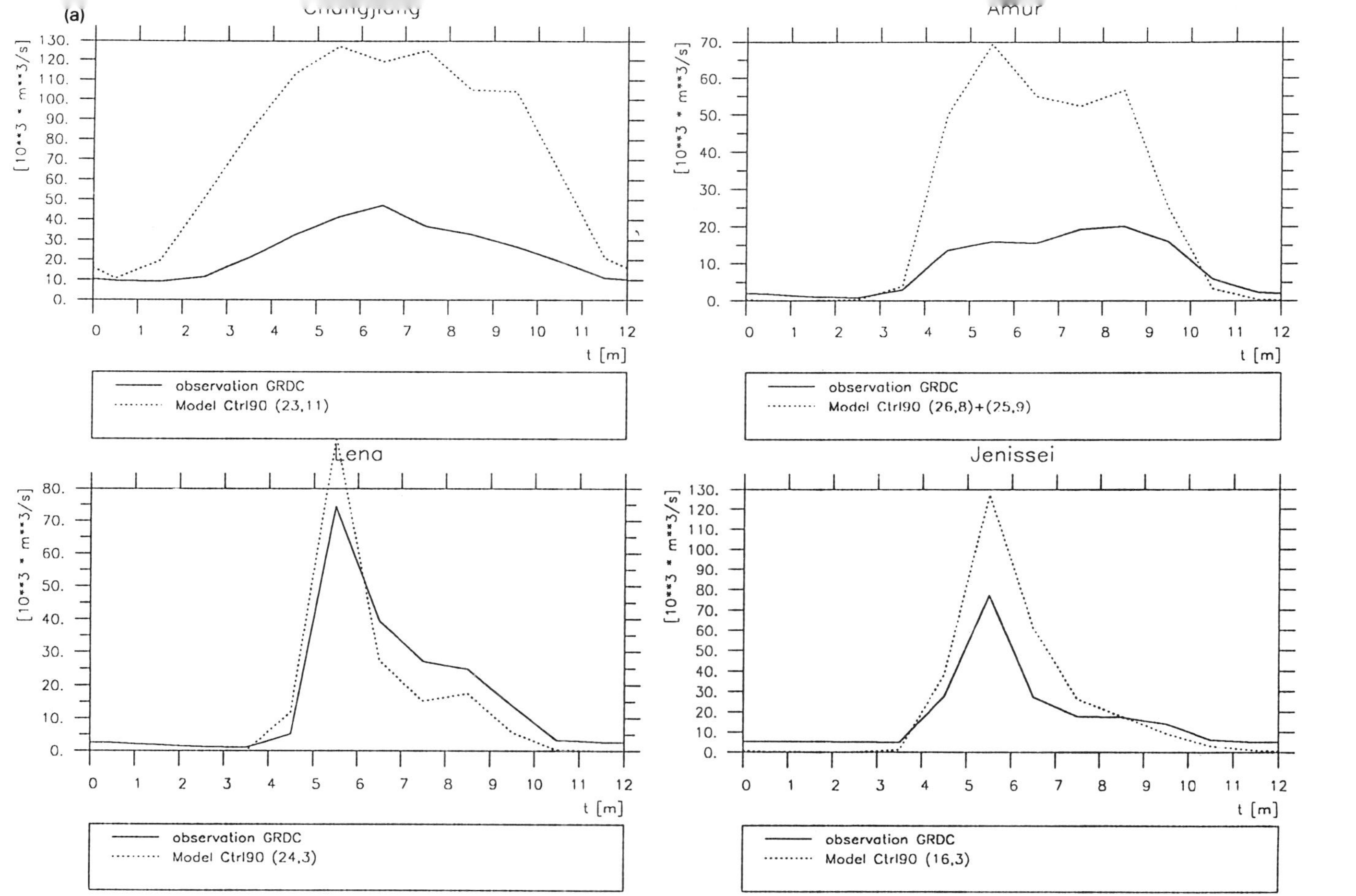

Fig. 11. Results from the 20-year integration with the ECHAM2-T21 model to which the river-runoff scheme by Sausen et al. (1991) was applied. Observations (full lines) are multi-year monthly averages which were kindly provided by GRDC, Koblenz, F.R.G. Model results are 20-year average discharges at the model grid points representative for the river (in brackets).

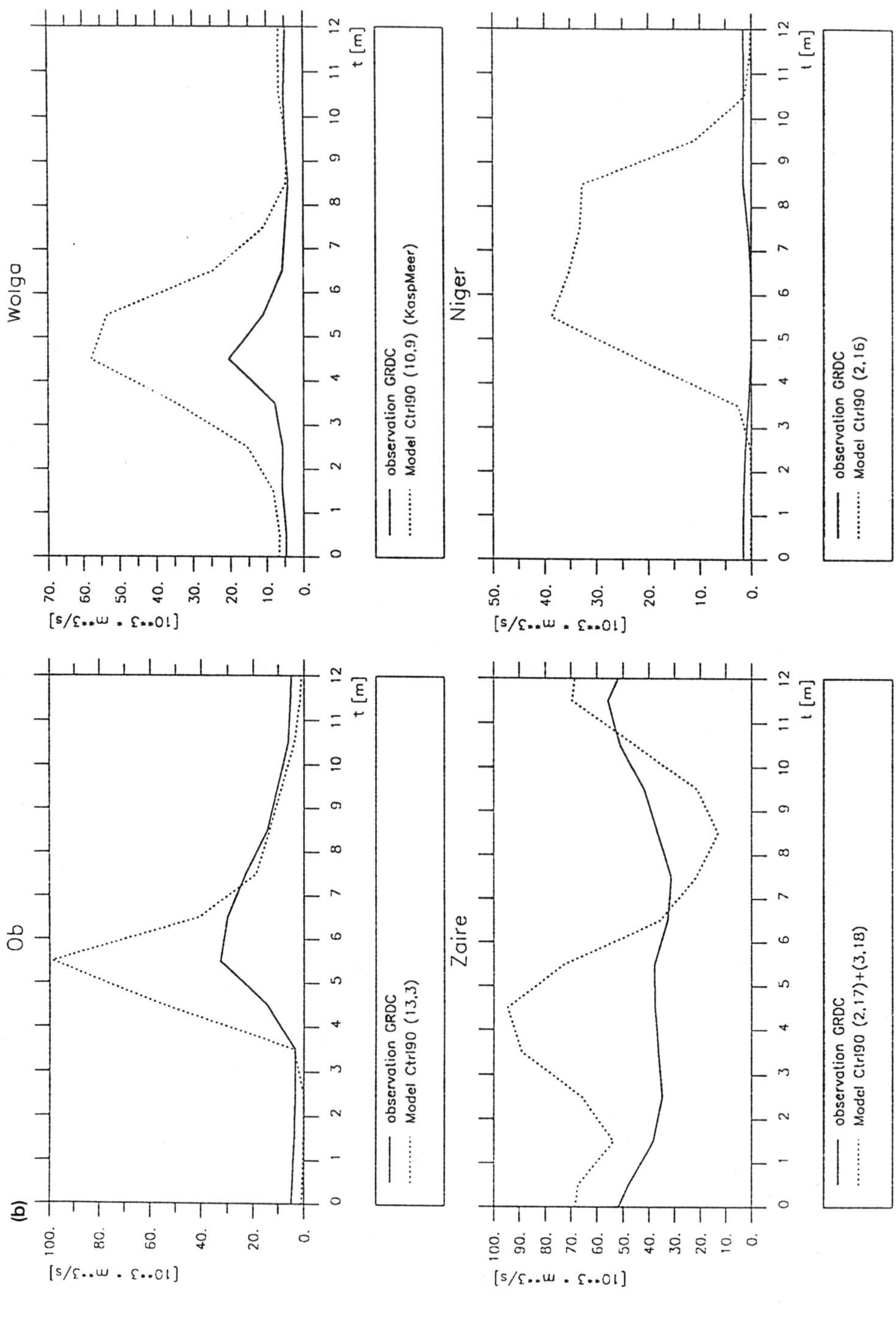

Fig. 11 (continued).

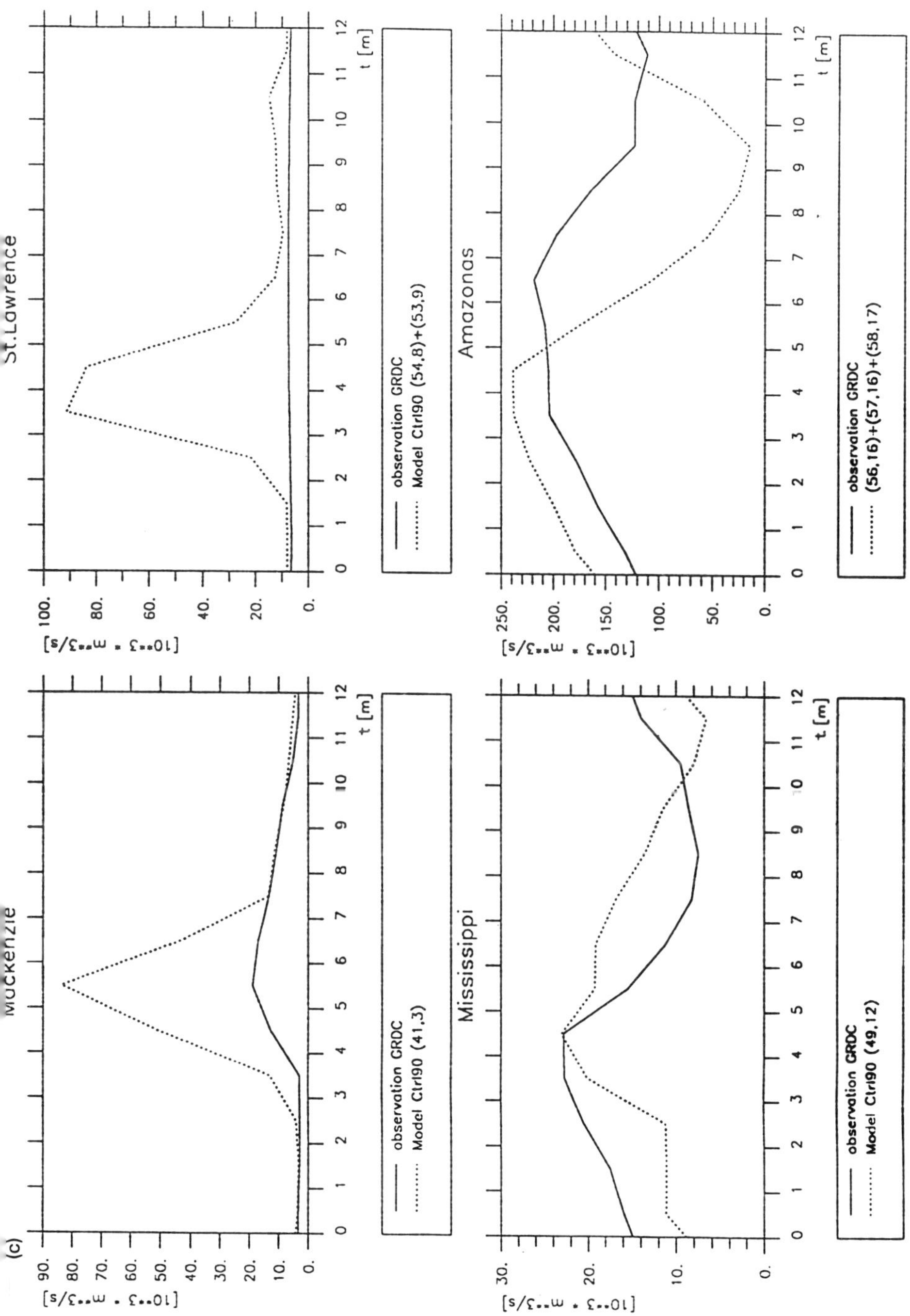

Fig. 11 (continued).

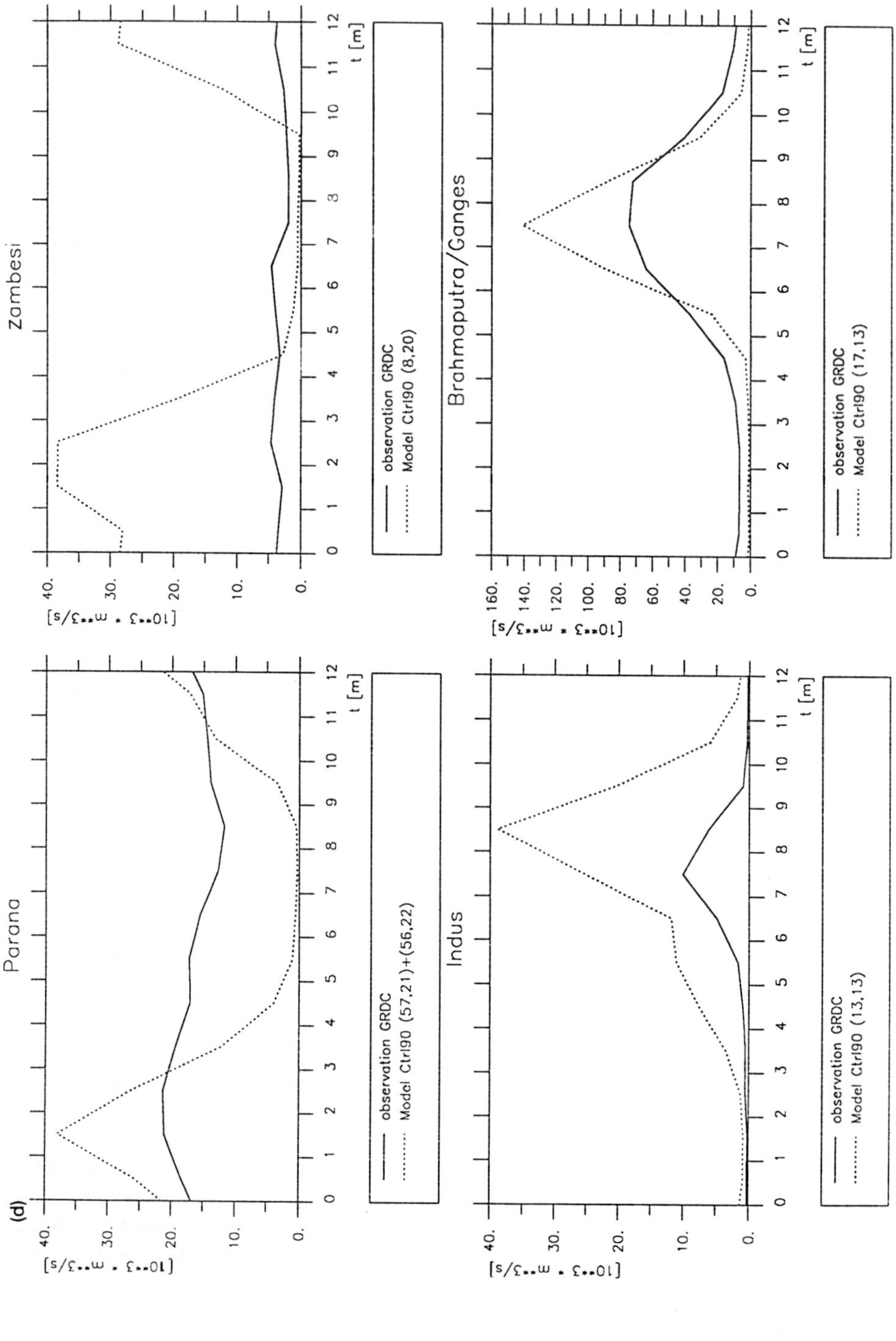

Fig. 11 (continued).

the model simulation, as a consequence, is in much better agreement with observed river gauge data.

5. Conclusions

The use of coupled ocean–atmosphere general circulation models for climate modelling, requires the computation of the time evolution of rainfall/runoff over the continents, which is one of the coupling parameters. The representation of this quantity in the OAGCM relies on two parameterisations: first, the specification of local runoff at a grid cell during a precipitation event and from the slowly acting drainage in the soil; second, the local runoff is collected in large river catchments and transferred to the river mouths, where it enters the ocean as freshwater input.

In the present paper we have discussed how the hydrological scheme for rainfall/runoff which was developed for the prediction of the water level of the river Arno, fulfils the requirements of the OAGCM. The heterogeneity of the land surface conditions over a grid area, determines the separation of rainfall between infiltration into the soil and surface runoff.

The scheme therefore is a first attempt at aggregating the physical processes at the point scale to the large scale of the OAGCM.

Applied in a 20-year integration with the ECHAM2 OAGCM, the scheme allows a consistent description of the various components of the hydrological cycle in the tropics, as well as, in the snow-covered regions of the Northern Hemisphere continents. We have attempted to validate qualitatively the model simulated hydrological cycle using independent sources of meteorological and hydrological data. In the future, gauge data from the largest rivers of the world will provide a particularly useful tool for independent verification of model results.

In its present configuration, the scheme is restricted to a uniform distribution of the precipitation over the grid cell. Future work will be directed at incorporating the small-scale features of the distribution of precipitation in space and time. It will also be necessary to expand the scheme, which is now restricted to one simple soil type, to soil types that vary for each grid cell. This should allow for a larger variability of the local runoff on the continental scale. At each grid point, the parameter b would then have to be computed as a function of soil type as well as orography.

The verification data and data for the initialisation of such extended schemes will have to be provided by the global experimentation and data collection exercises planned for the next decade.

Acknowledgement

Part of this work was carried out under the European Community Research Programme on Climatology, EPOCH project EV4C 0025I on Spatial Variability of Land Surface Processes.

References

Abbott, M. et al., 1986. An introduction to the European Hydrological System — Système Hydrologique Européen, SHE; 1. History and philosophy of a physically based distributed modelling system. J. Hydrol., 87: 45–59.

Baumgartner, A. and Reichel, E., 1975. The World Water Balance: Mean Annual Global Continental and Maritime Precipitation, Evaporation and Runoff. R. Oldenbourg, München–Wien.

Behr, H. and Dümenil, L., 1991. The snow climatology in decadal model simulations. Large-Scale Modelling Report No. 9, Meteorologisches Institut der Universität Hamburg, Hamburg, F.R.G.

Blondin, C., 1989. Research on land-surface parameterisation schemes at ECMWF. Proc. Workshop on Parameterisation of Fluxes over Land Surface, 24–26 October 1988, ECMWF, Reading, U.K.

Burnash, R.J.C., Ferral R.L. and McGuire, R.A., 1973. A general streamflow simulation system — conceptual modelling for digital computers. Report by the Joint Federal State River Forcasts Center, Sacramento, Calif.

Crawford, N.H. and Linsley, R.K., 1966. Digital simulation in hydrology, Stanford Watershed model IV. Tech. Rep. 39, Dep. Civil Eng. Stanford Univ., Stanford, Calif.

Deardorff, J.W., 1972. Parameterization of the planetary boundary layer for use in general circulation models. Mon. Weather Rev., 100: 93–106.

Fischer, G. (Editor), 1987. Climate simulations with the ECMWF T21 model in Hamburg. Large-Scale Atmospheric Modelling Report No. 1, Meteorologisches Institut der Universität Hamburg, Hamburg, F.R.G.

Jaeger, L., 1976. Monatskarten des Niederschlags für die ganze Erde. Berichte des Deutschen Wetterdienstes Nr. 139, Bd. 18, Offenbach/Main.

Legates, D.R., 1987. A climatology of global precipitation. Publ. Climatol. 40(1), Newark, Del., 85 pp.

Legates, D.R. and Willmott, C.J., 1990. Mean seasonal and spatial variability in gauge-corrected global precipitation. Int. J. Climatol., 10: 111–127.

Manabe, S., 1969. Climate and ocean circulation: I. The atmospheric circulation and the hydology of the earth's surface. Mon. Weather Rev., 97: 739–774.

Mintz, Y. and Serafini, V., 1981. Global field of soil water and land surface evapotranspiration. NASA Goddard Flight Center Tech. Memo. 83907, Res. Rev. 1980–1981, pp. 178–180.

Rocwood, D.M. and Nelson, M.L., 1966. Computer application to streamflow synthesis and reservoir regulation, IV Int. Conf. on Irrigation and Drainage.

Roeckner, E., Dümenil, L., Kirk, E., Lunkeit, F., Ponater, M., Rockel, B., Sausen R. and Schlese, U., 1989. The Hamburg version of the ECMWF model. In: G.J. Boer (Editor), Research Activities in Atmospheric and Oceanic Modelling. CAS/JSC Working Group on Numerical Experimentation, 13, WMO/TD-No. 332, pp. 7.1–7.4.

Roeckner, E., Arpe, K., Bengtsson, L., Brinkop, S., Dümenil, L., Kirk, E., Lunkeit, F., Esch, M., Ponater, M., Rockel, B., Sausen, R., Schlese, U., Schubert, S. and Windelband, M., 1992. Simulation of the present-day climate with the ECHAM model: impact of model physics and resolution. Max Planck Institute Report No. 95.

Sausen, R., Schubert S. and Dümenil, L., 1991. A model of the river-runoff for use in coupled atmosphere–ocean models. Large-Scale Atmospheric Modelling Report No. 9, Meteorologisches Institut der Universität Hamburg, Hamburg, F.R.G.

Sittner, W.T., Schaus C.E. and Monro, J.C., 1969. Continuous hydrograph synthesis with an API-type hydrologic model. Water Resour. Res., 5(5): 1007–1022.

Sugawara, M., Ozaki, E., Watanabe, I. and Katsuyama, Y., 1976. Tank Model and its application to Bird Creek, Wollombi Brook, Bihin River, Sanaga River, and Nam Mune. Research Notes of the National Centre for Disaster Prevention, No. 11, Tokyo.

Todini, E., 1988. Il modello afflussi deflussi del fiume Arno. Relazione Generale dello studio per conto della Regione Toscana, University of Bologna.

Warrilow, D.A., Sangster A.B. and Slingo, A., 1986. Modelling of land surface processes and their influence on European climate. DCTN 38, U.K. Meteorological Office, Bracknell, U.K.

Zhao, R.J., 1977. Flood forecasting method for humid regions of China. East China College of Hydraulic Engineering, Nanjing, China.

Chapter 10

Transient response of a coupled ocean–atmosphere–land surface model to increasing atmospheric carbon dioxide

S. MANABE, R.J. STOUFFER, M.J. SPELMAN and K. BRYAN

Abstract

This study investigates the response of a climate model to a gradual increase of atmospheric carbon dioxide. The model is a general circulation model of the coupled ocean–atmosphere–land surface system with a global computational domain, smoothed geography, and seasonal variation of insolation. It is found that the simulated increase of sea surface temperature is very slow over the northern North Atlantic and the Circumpolar Ocean of the Southern Hemisphere where the vertical mixing of water penetrates very deeply and the rate of deep water formation is relatively fast. With the exception of these two regions identified above, the distribution of the change in surface temperature of the model is qualitatively similar to the equilibrium response of an atmospheric–mixed layer ocean model, which has been the subject of many previous studies. In most of the Northern Hemisphere, the seasonal dependence of surface air temperature change is also similar to the equilibrium response. For example, the temperature increase is at a maximum over the Arctic Ocean and its surroundings in the late fall and winter, whereas it is at a minimum in summer. However, the increase of surface air temperature and its seasonal variation is very small in the Circumpolar Ocean of the Southern Hemisphere and the northern North Atlantic

The increase of atmospheric carbon dioxide affects not only the thermal structure of the coupled model but also its hydrologic cycle. For example, the global mean rates of both precipitation and evaporation increase. The increase in evaporation rate is particularly large in low latitudes and decreases with increasing latitudes. On the other hand, the increase in the precipitation rate is substantial in high latitudes due to the increased penetration of warm, moisture-rich air into high latitudes. Thus, the rate of runoff in the subarctic basin increases markedly.

In qualitative agreement with the results of equilibrium response studies, soil moisture is reduced in summer over extensive regions of the middle and high latitudes, such as the North American Great Plains, Western Europe, Northern Canada, and Siberia.

1. Introduction

The interaction among the atmosphere, land surface and ocean is the most fundamental process that controls the future climate change induced by greenhouse gases.

The transient response of climate to an increase of atmospheric carbon dioxide has been the subject of many studies: Hoffert et al. (1980), Bryan et al. (1982), Thompson and Schneider (1982), Hansen et al. (1984), Spelman and Manabe (1984), Harvey and Schneider (1985), Schlesinger et al. (1985), Hansen et al. (1988), Bryan et al. (1988), Washington and Meehl (1989), Stouffer et al. (1989), Manabe et al. (1990), and Manabe et al. (1991).

This presentation describes the response of a coupled ocean–atmosphere–land surface model developed at GFDL to a gradual increase of atmospheric carbon dioxide. It summarizes some of the results in three recently published papers (Stouffer et al., 1989; Manabe et al., 1991, 1992). They represent the current state-of-the-art in predicting further climate change induced by greenhouse gases in the atmosphere.

In response to an increase of atmospheric carbon dioxide, for example, the infrared opacity of the atmosphere increases, thereby raising the effective source of emission for the outgoing terrestrial radiation to a higher level with lower temperature. The reduction of the effective emission temperature results in the warming of both the convective troposphere and earths's surface which are coupled closely to each other by the efficient boundary layer heat exchange. The rise of tropospheric temperature, in turn, induces an increase of the moisture content of air, further enhancing the infrared opacity of the atmosphere and raising the temperature of the troposphere–surface system.

The warming of the coupled system is reduced by the effective thermal inertia of the earth's surface which is essentially controlled by the vertical mixing of heat in the oceans. This study investigates how such vertical mixing of heat is achieved, resulting in the delay of the greenhouse warming in the coupled troposphere–ocean–continental surface system.

2. Model structure

The model consists of general circulation models (GCM) of the atmosphere and oceans, and a simple model of the continental surface that involves the budgets of heat and water. It is a global model and includes realistic geography. The atmospheric component of the model used here is very similar to the model used by Manabe and Hahn (1981). The seasonally varying insolation at the top of the model atmosphere is taken into account in calculating the transfer of solar radiation through the atmosphere.

Precipitation is computed whenever supersaturation is indicated by the prognostic equation for water vapor. It is identified as snowfall when the air temperature

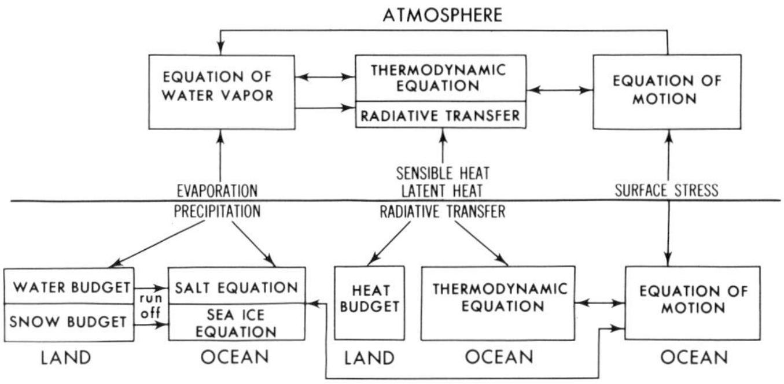

Fig. 1. Box diagram illustrating the coupling between the oceanic, atmospheric, and land surface components of the model.

near the surface falls below freezing; otherwise it is identified as rain. The moist convective processes are parameterized by a moist convective adjustment scheme (Manabe et al., 1965). Cloud cover is predicted whenever the relative humidity exceeds a certain critical value, which is 99% in this case. Otherwise, clear sky is predicted. The distribution of cloud cover thus determined is used for the computation of solar and terrestrial radiation.

A change in snow depth is computed as the net contribution from snowfall, sublimation, and snowmelt that is determined from the requirement of surface heat balance. The budget of soil moisture is computed by the so-called bucket method (Manabe, 1969). For simplicity, it is assumed that the soil can hold 15 cm of liquid water. When soil is not saturated with water, the change in soil moisture is predicted as the net contribution of rainfall, evaporation, and snowmelt. If the bucket is full, the excess water is regarded as runoff. The rate of evaporation from the soil surface is determined as a function of the water content of the "bucket" and potential evaporation.

The atmospheric GCM uses a spectral transform method in which the horizontal distributions of predicted variables are represented by spherical harmonics (15 associated Legendre functions for each of 15 Fourier components) and grid-point values (Gordon and Stern, 1982). For vertical finite-difference calculations, nine unequally spaced levels are chosen. The ocean GCM uses a full finite difference technique and has a regular grid system with 4.5° × 3.75° (latitude × longitude) spacing and 12 vertical levels (Bryan and Lewis, 1979). The atmospheric and oceanic components of the model interact with each other continuously through the exchange of heat, water, and momentum fluxes as illustrated by Figure 1.

3. Numerical experiments

To study the response of a coupled ocean–atmosphere–land surface model to a gradual increase of atmospheric carbon dioxide, two 100-year integrations were performed. Starting from an initial condition in a quasi-equilibrium state, the standard time integration of the coupled model (i.e., S integration) was performed with a normal concentration of atmospheric carbon dioxide. In the other integration the CO_2 concentration was increased by 1%/year (compounded). Hereafter, the latter integration is identified as the G integration. The rate of 1%/year is chosen because the total CO_2-equivalent radiative forcing of various greenhouse gases other than water vapor is currently increasing approximately at this rate (Ramanathan et al., 1985).

The initial conditions for both integrations have realistic seasonal and geographical distributions of surface temperature, surface salinity, and sea ice, with which both the atmospheric and oceanic components of the model are nearly in equilibrium. This quasi-equilibrium condition was obtained by separate time integrations of these two components of the model using observed surface conditions. The convergence of the oceanic component towards equilibrium was accelerated by the method described in Bryan (1984).

To avoid systematic climatic drift caused by the bias of the model during the S integration, the oceanic surface fluxes of heat and water are adjusted seasonally and geographically, but not interannually, using the method described by Manabe and Stouffer (1988), and Manabe et al. (1991). Identical adjustments are applied to the G integration defined above. In contrast to recent studies (Schlesinger and Jiang, 1988; Washington and Meehl, 1989), no systematic trends are evident during the S integration, indicating that the initial condition is close to stable equilibrium and the flux adjustments are working as expected.

4. Temperature change

4.1. Annual mean response

Figure 2 shows the latitude–time distribution of the CO_2-induced change of zonally averaged, decadal-mean surface air temperature during the 100-year experiment. In the Northern Hemisphere the warming of surface air increases with latitude, partly because of the poleward retreat of both snow cover and sea ice, which have high surface albedo. The relatively large warming in high northern latitudes is essentially confined to the lower troposphere because of the stable stratification. This is indicated in Figure 3, which shows the difference in zonal-mean temperature between the two integrations averaged over the 60- to 80th-year period of the experiment. The ocean warming is essentially confined to the upper layer, except near Antarctica and around 70°N. It also increases with latitude in the Northern Hemisphere, but becomes small in the Arctic Ocean, which is covered by sea ice.

According to Figure 4 the simulated response of surface air temperature is very slow over the Circumpolar Ocean of the Southern Hemisphere and the northern North Atlantic. As explained in this and earlier studies, the vertical mixing of water extends very deeply in these regions through the combined effect of thermal

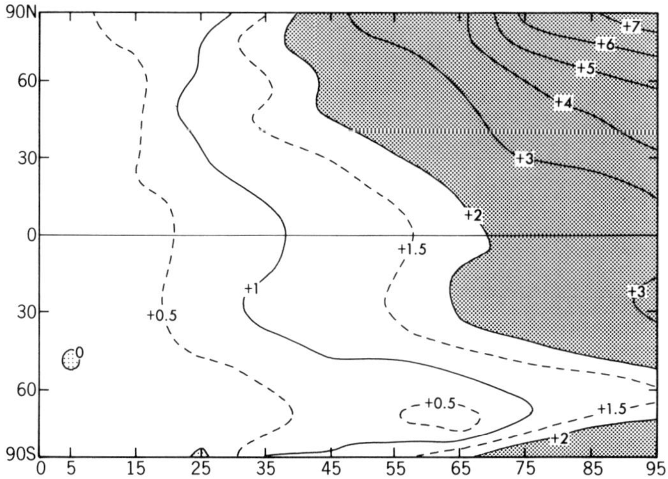

Fig. 2. Latitude–time distribution of the difference in zonally averaged decadal-mean surface air temperature (°C) between the two integrations.

advection by ocean currents and vertical subgrid-scale mixing including convection. Because of this deep vertical mixing, heat trapped by the increasing atmospheric carbon dioxide spreads deeply over a thick layer of water with large heat capacity, thereby making the warming of surface temperature very small in these oceanic regions.

In both the northern North Atlantic and the Circumpolar Ocean of the Southern Hemisphere, there are other physical mechanisms that reduce the CO_2-induced increase of surface air temperature. Refer to the study of Manabe et al. (1991) for the elucidation of these processes.

Based upon the results from current climate models, the WMO/UNEP Intergovernmental Panel on Climate Change (1990) predicted the rate of increase of global mean temperature during the next century of about 0.3°C per decade (with an uncertainty range of 0.2°C to 0.5° per decade). The results from the present model belong to the upper half of this estimate (see Figure 5).

One of the major factors which have an effect upon climate and its sensitivity is cloud cover. Cloud reflects a large fraction of solar radiation and traps outgoing terrestrial radiation, thereby exerting a large influence upon the radiation budget of this planet. Our inability to reliably model the change of cloud cover and its effect upon climate accounts partly for the large uncertainty in our current estimate of future climate change induced by greenhouse gases.

4.2. Seasonal dependence

In response to a gradual increase of atmospheric carbon dioxide, the increase of surface air temperature is at a maximum over the Arctic Ocean and its surroundings in the late fall and winter. On the other hand, the Arctic warming is at a minimum

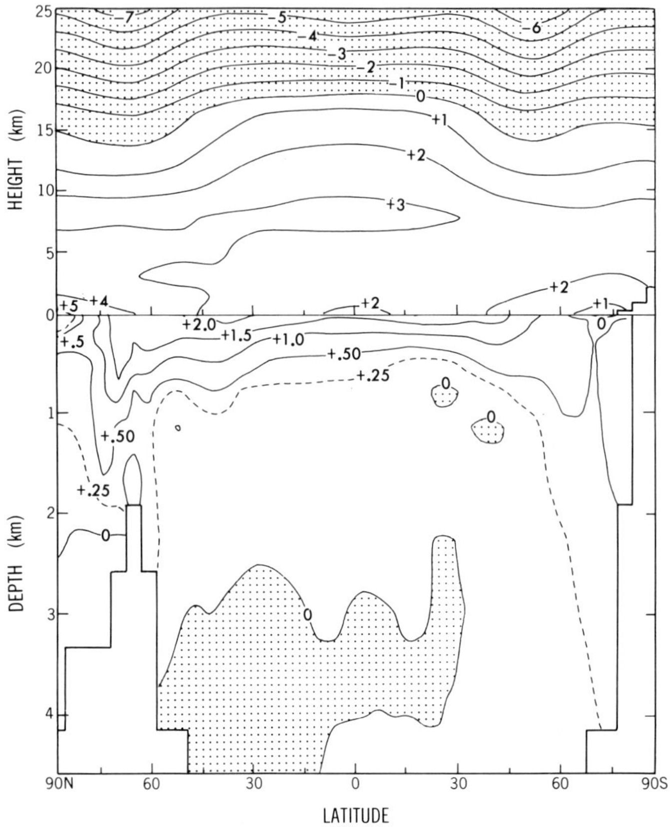

Fig. 3. CO_2-induced change of zonal mean temperature averaged over the 60th–80th year period of the experiment.

in summer (Figure 6). The mechanism responsible for the seasonal dependence of the surface air temperature change described above is essentially similar to the results from an equilibrium experiment which was conducted earlier (Manabe and Stouffer 1979, 1980) with an atmosphere–mixed layer ocean model. Because of the existence of a halocline with very stable stratification, the heat exchange between the mixed layer and deeper ocean is very small in the Arctic Ocean. Therefore, it is not surprising that the seasonal dependence of surface air temperature change in the coupled model is qualitatively similar to the response of an atmosphere–mixed layer ocean model in which the heat flux at the bottom of the mixed layer is unchanged.

The results from the present study indicate that, during the fall and winter, the large increase of surface air temperature is achieved by the enhanced upward heat conduction from the mixed layer ocean to the atmosphere through thinner sea ice. On the other hand, the downward flux of terrestrial radiation increases in response to the increase of greenhouse gases (i.e., CO_2 and H_2O) in the atmosphere, and the

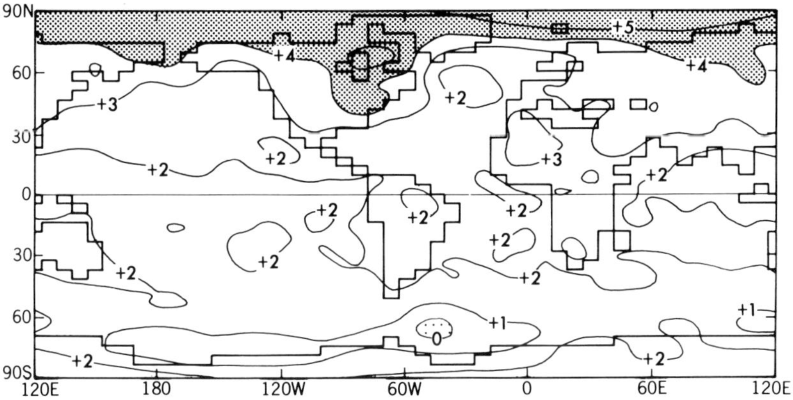

Fig. 4. The transient response of the surface air temperature (°C) of the coupled ocean–atmosphere model to the 1%/year increase of atmospheric carbon dioxide. The response is the difference between the G and the S integrations averaged over the 60th–80th year period when the CO_2 concentration is doubled.

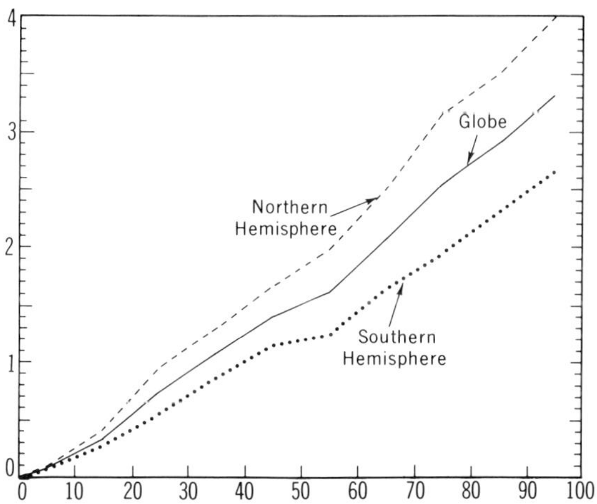

Fig. 5. The temporal variation of the difference in area averaged, decadal-mean surface air temperature (°C) between the integrations G and S. Solid, dashed and dotted lines indicate the differences averaged over the globe and Northern and Southern Hemispheres, respectively.

reflection of solar radiation by the surface decreases due to the reduced coverage of sea ice during the warm season, enhancing the absorption of radiative energy by the mixed layer ocean. However, the temperature of the sea surface and overlying air hardly changes in summer because they are anchored near the freezing point due to the melting of sea ice. The additional heat, which is thus stored in the mixed layer ocean, is lost to the atmosphere during the cold season (as described above) due to the enhanced heat conduction through thinner sea ice.

In sharp contrast to the situation in the Arctic Ocean, the increase of surface air temperature and its seasonal variation in the Circumpolar Ocean of the Southern

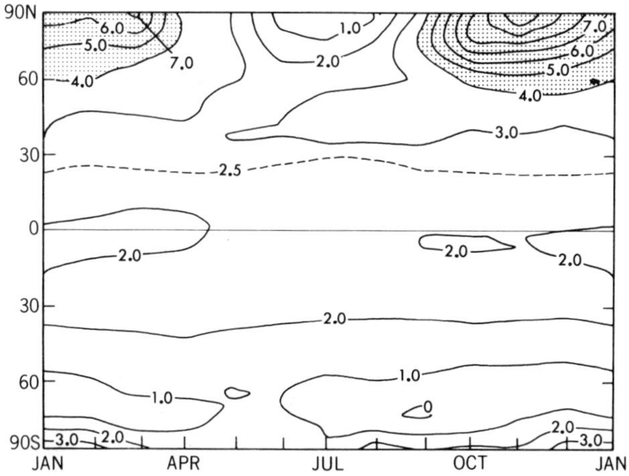

Fig. 6. Seasonal and latitudinal variation of the difference in zonal mean surface air temperature (°C) between the G and S integrations averaged over the 60th–80th year period when the CO_2 concentration is doubled.

Hemisphere are very small due to the deep vertical mixing of water discussed earlier. Thus, the maximum warming during the cold season does not materialize in the Circumpolar Ocean. This is quite different from the results of the earlier equilibrium experiments which indicate the maximum warming during the cold season.

5. Hydrologic change

5.1 Annual mean response

Figure 7 illustrates the CO_2-induced change in the zonal mean rates of precipitation and evaporation averaged over the 60- to 80th-year period of the experiment. This figure indicates that, as noted in many previous studies (e.g., Manabe and Wetherald, 1980), the rates of both precipitation and evaporation increase resulting from an increase of atmospheric carbon dioxide. However, the latitudinal profile of the CO_2-induced increase in precipitation rate is quite different from the corresponding profile of evaporation. Specifically, the change in evaporation rate decreases with increasing latitudes and is very small in high latitudes. On the other hand, the enrichment of the water vapor content of air accompanying the atmospheric warming enhances the poleward moisture transport, thereby causing a marked increase in the precipitation rate in high latitudes. Accordingly, the excess of precipitation over evaporation also increases substantially in these latitudes and contributes to the marked increase in river runoff in the subarctic regions.

It is notable that in the present experiment the precipitation rate increases substantially in high latitudes of the Southern Hemisphere where the CO_2-induced

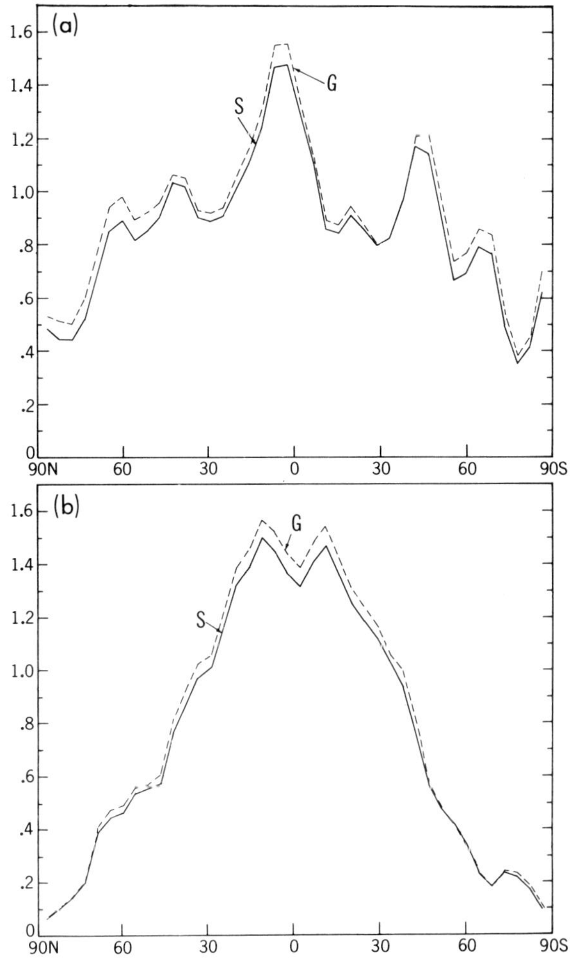

Fig. 7. The latitudinal profiles of zonal mean rates of precipitation (a), and evaporation (b), averaged over the 60th–80th year period of the experiment. Solid line: standard integration. Dashed line: CO_2-growth integration.

surface warming is small. As indicated in Figure 3, significant mid-tropospheric warming occurs in high latitudes of the Southern Hemisphere enhancing the penetration of warm, moisture-rich air towards the Antarctic Continent. Thus, the large excess of precipitation over evaporation should contribute to the accretion of the ice sheet over the continent.

The CO_2-induced changes in the rates of precipitation and evaporation described above increase the excess of precipitation over evaporation in high latitudes where salinity is reduced in the upper layer of the model ocean. For example, the capping of the Atlantic Ocean by water of relatively low salinity weakens the thermohaline circulation, moderating the CO_2-induced warming of the near surface layer of the atmosphere in the northern North Atlantic. This is a topic of extensive discussion in our recent studies (Manabe et al., 1990, 1991).

5.2. Seasonal dependence

The gradual increase of atmospheric carbon dioxide also affects the seasonal variation of land surface hydrology. For example, in response to the CO_2 increase, soil moisture is reduced during the June–July–August period over most of the continental regions with the notable exception of the Indian Subcontinent where precipitation and soil moisture increase (Figure 8a). The summer reduction of soil moisture in the Northern Hemisphere is relatively large over the middle latitude belt, i.e., the great plains of the North American Continent, Eastern China, Southern Europe and Siberia in qualitative agreement with results from equilibrium experiments conducted in earlier studies by use of an atmosphere–

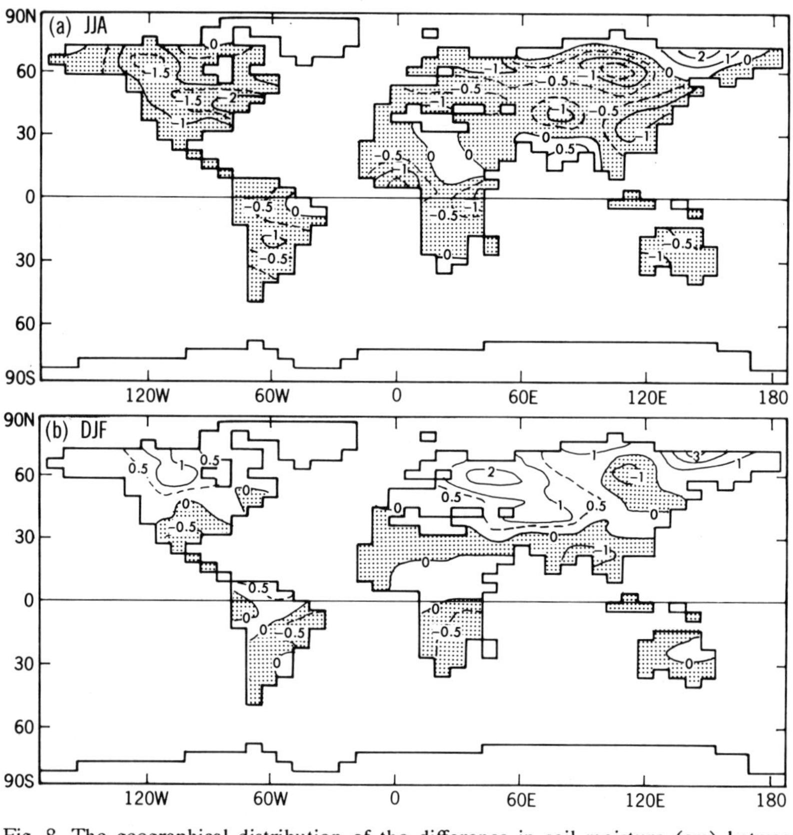

Fig. 8. The geographical distribution of the difference in soil moisture (cm) between the G and S integrations averaged over the 60th–80th year period when CO_2 concentration is doubled. (a) June–July–August. (b) December–January–February. Note that the field capacity of soil moisture is assumed to be 15 cm in the present experiment.

mixed layer ocean model. The summer drying is mainly caused by an earlier termination of snowmelt and rainy periods in spring, and thus, an earlier onset of the normal spring-to-summer reduction of soil moisture (see, for example, Manabe and Wetherald, 1987).

During the December–January–February period, soil moisture increases in middle and high latitudes of the Northern Hemisphere (Figure 8b). For example, the increase is relatively large over Western Europe and most of Canada. On the other hand, it is reduced in the subtropics particularly over Southeast Asia and Mexico. The reduction of soil moisture is also notable in the subtropical portion of the continents in the Southern Hemisphere in winter (i.e., June–July–August period) although the regions of relatively large reduction change with time as influenced by the natural variability of the model climate.

The broad features of soil moisture change and its seasonal variation in the present experiment are qualitatively similar to the change in the equilibrium experiments conducted in many earlier studies. The mechanisms involved in soil moisture change are also similar between the transient and equilibrium experiments. One can detect, however, another mechanism which reduces slightly soil moisture and runoff over the continents in middle and high latitudes of the Northern Hemisphere in the transient experiment. Because of the thermal inertia of the ocean, the increase of sea surface temperature is reduced (or delayed) substantially in middle and high northern latitudes. Thus, the increase in oceanic evaporation, and accordingly, that of moisture transport from the oceans to continents are less than what they would be without the delay in the warming of the sea surface, resulting in the reduction of precipitation, soil moisture and runoff over the continents polewards of 30° latitude in the Northern Hemisphere.

6. Summary and concluding remarks

It has been shown that the rise of surface air temperature in response to the gradual increase of atmospheric carbon dioxide increases with increasing latitude in the Northern Hemisphere. Over the Arctic Ocean and the surrounding region where sea ice decreases markedly, the warming is particularly large in winter but is relatively small in summer. Owing to the oceanic influences, the warming is delayed substantially over the northern North Atlantic and modestly in western Europe. In sharp contrast to the situation in the Northern Hemisphere described here, the warming decreases with increasing latitude in the Southern Hemisphere. In particular, surface air temperature hardly changes in the vicinity of the Antarctic Continent due to the large effective thermal inertia of the Circumpolar Ocean with deep vertical mixing. It is also surprising that the thickness of sea ice increases significantly in both the Weddel and Ross Seas. These results indicate that the suggested collapse of the West Antarctic Ice Sheet may be a very remote possibility.

The numerical experiment conducted here indicates that the oceans can exert a profound influence upon the distribution of the CO_2-induced change of surface air temperature, although many of the features of the equilibrium response obtained from the previous studies still remain valid in most of the Northern Hemisphere and lower latitudes off the Southern Hemisphere. As discussed above, the effective thermal inertia of the oceans as influenced by oceanic circulation is particularly large over the northern North Atlantic and its surroundings and over the Circumpolar Ocean of the Southern Hemisphere where the CO_2-induced temperature change is small.

Obviously the credibility of the present study depends critically upon the ability of the present model to mimic the behavior of the actual ocean–atmosphere–land surface system. In this connection, it is encouraging that Dixon et al. (pers. commun.) recently succeeded in reproducing the downward spreading of chlorofluorocarbons in the Circumpolar Ocean of the Southern Hemisphere by employing the coupled model used in the present study. However, the adjustments in the surface fluxes of heat and water, which are needed to simulate the condition of the oceanic surface, are as large as those fluxes themselves. For example, the model tends to overestimate the precipitation in high latitudes and is partly responsible for the relatively large water flux adjustment needed there. It is possible that such a model bias results in the overestimation of the CO_2-induced change of precipitation, and accordingly, the change of near-surface salinity and its effect upon thermohaline circulation and convective overturning.

The increase of a greenhouse gas affects not only the thermal structure of the coupled system but also the hydrologic cycle. For example, the global mean rates of both precipitation and evaporation increase. In high latitudes of the Northern Hemisphere the increase in the rate of precipitation far exceeds that of evaporation, markedly increasing the rate of runoff and reducing the surface salinity in the Arctic and surrounding oceans. This capping of the oceanic surface by relatively fresh water weakens the thermohaline circulation, reducing the surface warming in the northern North Atlantic.

It has been noted that soil moisture is reduced in summer over extensive mid-continental regions of both Eurasian and North American Continents of the model. Thus, it is likely that summer drought may become more frequent as the greenhouse warming intensifies. One should keep in mind, however, that the change of soil moisture over the continental surface critically depends upon the value of soil moisture in the control integration. For example, the summer reduction of soil moisture in an equilibrium CO_2 doubling experiment conducted earlier failed to materialize in those regions where soil moisture is very small in the control integration. Unfortunately, the simulations of precipitation and soil moisture are far from satisfactory. Thus, one should not take too literally the geographical details of the changes of climate obtained here. A more realistic simulation of precipitation is required for a reliable estimate of soil moisture change.

The present study corroborates the projection of the WMO/UNEP Intergovernmental Panel on Climate Change (1990) that the rate of the increase in global mean surface air temperature may be a few tenths of a degree centigrade per century. Unfortunately, the projection is subject to a large uncertainty because of

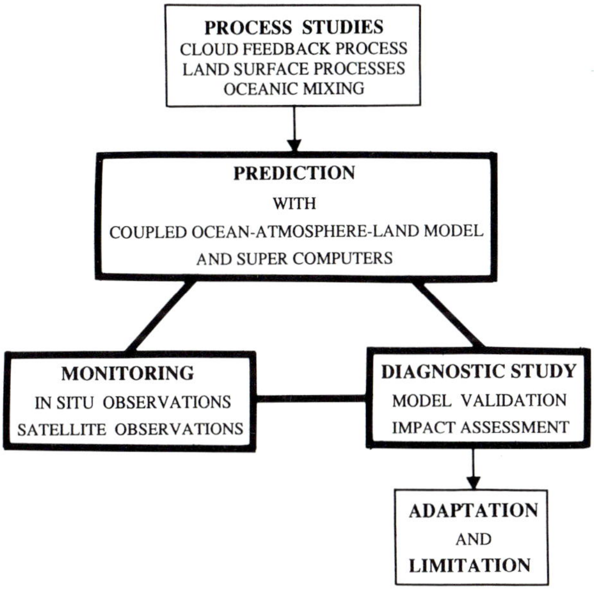

Fig. 9. Diagram illustrating the triad strategy for the prediction of interdecadal climate change.

our inability to model various processes that control the future climate change. It is therefore necessary to improve various components of the climate model, such as the cloud feedback and land surface processes. In addition, it is essential to carefully assess the model predictions of future climate change in light of the results obtained by monitoring the actual climate changes and the factors causing these changes. The agreement between the predicted and observed climate should enhance our confidence in model prediction. A comprehensive triad strategy for predicting future climate is illustrated by a box diagram in Figure 9. It involves the monitoring of the coupled ocean–atmosphere–land surface system by in situ and remote sensing, the prediction of future climate by state-of-the-art models, and in-depth analysis of predicted and monitored climate. The insight gained by this comprehensive effort is indispensable not only for enhancing our confidence in model prediction of future climate, but also for adapting to future climate change and limiting the emission of greenhouse gases.

References

Bryan, K., 1984. Accelerating the convergence to equilibrium of ocean–climate models. J. Phys. Oceanogr., 14: 666–673.

Bryan, K. and Lewis, L.J., 1979. A water mass model of the world ocean. J. Geophys. Res., 84(C5): 2503–2512.

Bryan, K., Komro, F.G., Manabe S. and Spelman, M.J., 1982. Interhemispheric asymmetry in the transient response to increasing atmospheric carbon dioxide. Science, 215: 56–58.

Bryan, K., Manabe S. and Spelman, M.J., 1988. Interhemispheric asymmetry in the transient response of a coupled ocean–atmospheric model to a CO_2 forcing. J. Phys. Oceanogr., 18: 851–867.

Cox, M.D., 1987. Isopycnal diffusion in a Z-coordinate ocean model. Ocean Modeling, 74: 1–5.

Gordon, C.T. and Stern, W., 1982. A description of the GFDL Global Spectral Model. Mon. Weather Rev., 110: 625–644.

Hansen, J., Lacis, A., Rind, D., Russell G. and Stone, P., 1984. Climate sensitivity: Analysis of feedback mechanisms. In: J.E. Hansen and T. Takahashi (Editors), Climate Processes and Climate Sensitivity. Geophysical Monograph 29, Maurice Ewing Vol. 5, AGU, Washington, D.C., pp. 130–163.

Hansen, J., Fung, I., Lacis, A., Rind, D., Lebedeff, S., Ruedy R. and Russell, G., 1988. Global climate changes as forecast by the Goddard Institute for Space Sciences three-dimensional model. J. Geophys. Res., 93: 9341–9364.

Harvey, L.D.D. and Schneider, S.H., 1985. Transient climate response to external forcing on 10°–104 year time scales. J. Geophys. Res., 90: 2191–2205.

Hoffert, M.K., Callegari A.J. and Hsieh, C.-T., 1980. The role of deep sea storage in the secular response to climate forcing. J. Geophys. Res., 85: 6667–6679.

Manabe, S., 1969. Climate and the ocean circulation: I. The atmospheric circulation and the hydrology of the earth's surface. Mon. Weather Rev., 97: 739–774.

Manabe, S. and Hahn, D.G., 1981. Simulation of atmospheric variability. Mon. Weather Rev., 109: 2260–2286.

Manabe, S. and Stouffer, R.J., 1979. A CO_2-climate sensitivity study with a mathematical model of the global climate. Nature, 282: 491–493.

Manabe, S. and Stouffer, R.J., 1980. Sensitivity of a climate model to an increase of CO_2 concentration in the atmosphere. J. Geophys. Res., 85(C10): 5529–5554.

Manabe, S. and Stouffer, R.J., 1988. Two stable equilibria of a coupled ocean–atmosphere model. J. Climate, 1: 841–866.

Manabe, S. and Wetherald, R.T., 1980. On the distribution of climate change resulting from an increase in CO_2 content of the atmosphere. J. Atmos. Sci., 37: 99–118.

Manabe, S. and Wetherald, R.T., 1987. Large scale changes of soil wetness induced by an increase in atmospheric carbon dioxide. J. Atmos. Sci., 44: 1211–1245.

Manabe, S., Smagorinsky, J. and Strickler, R.F., 1965. Simulated climatology of a general circulation model with a hydrologic cycle. Mon. Weather Rev., 93: 769–798.

Manabe, S., Bryan K. and Spelman, M.J., 1990. Transient response of a global ocean–atmosphere model to a doubling of atmospheric carbon dioxide. J. Phys. Oceanogr., 20: 722–749.

Manabe, S., Stouffer, R.J., Spelman M.J. and Bryan, K., 1991. Transient responses of a coupled ocean–atmosphere model to gradual changes of atmospheric CO_2: Part I. Annual mean response. J. Climate, 4: 785–818.

Manabe, S., Spelman M.J. and Stouffer, R.J., 1992. Transient responses of a coupled ocean–atmosphere model to gradual changes of atmospheric CO_2. Part II: Seasonal Response. J. Climate, 5: 105–126.

Meehl, G.A. and Washington, W.M., 1988. A comparison of soil moisture sensitivity in two global climate models. J. Atmos. Sci., 45: 1476–1492.

Mitchell, J.F.B. and Warrilow, D.A., 1987. Summer dryness in northern mid-latitudes due to increased CO_2. Nature, 330: 238–240.

Ramanathan, V., Cicerone, R.J., Singh H.B. and Kiehl, J.T., 1985. Trace gas trends and their potential role in climate change. J. Geophys. Res., 90: 5547–5566.

Redi, M.H., 1982. Oceanographic isopycnal mixing by coordinate rotation. J. Phys. Oceanogr., 12: 1154–1158.

Schlesinger, M.E. and Jiang, X., 1988. The transport of CO_2-induced warming into the ocean: An analysis of simulations by the OSU coupled atmosphere–ocean general circulation model. Clim. Dyn., 3: 1–17.

Schlesinger, M.E., Gates W.L. and Han, Y.J., 1985. The role of the ocean in CO_2-induced climatic warming: Preliminary results from the OSU coupled atmosphere–ocean GCM. In: J.C.J. Nihoul (Editor), Coupled Ocean–Atmosphere Models. Elsevier, New York, N.Y., pp. 447–478.

Spelman, M.J. and Manabe, S., 1984. Influence of oceanic heat transport upon the sensitivity of a model climate. J. Geophys. Res., 8 9: 571–586.

Stouffer, R.J., Manabe S. and Bryan, K., 1989. Interhemispheric asymmetry in climate response to a gradual increase of atmospheric CO_2. Nature, 342: 660–662.

Thompson, S.L. and Schneider, S.H., 1982. Carbon dioxide and climate: The importance of realistic geography in estimating the transient temperature response. Science, 217: 1031–1033.

Washington, W.M. and Meehl, G.A., 1989. Climate sensitivity due to increased CO_2: Experiment with a coupled atmosphere and ocean gen eral circulation model. Clim. Dyn., 4: 1–38.

Wetherald, R.T. and Manabe, S., 1988. Cloud feedback processes in a general circulation model. J. Atmos. Sci., 45,: 1397–1415.

WMO/UNEP Intergovernmental Panel on Climate Change, 1990. Scientific assessment of climate change: The policy maker's summary of the report of the working group I. WMO/UNEP, Geneva, 26 pp.

Part C. The past and the future

Chapter 11

Quantitative hydrology in Scandinavia in the 18th century

LARS GOTTSCHALK

Abstract

The honour of first using quantitative methods in the analysis of the hydrological cycle belongs to Pierre Perrault, as pointed out by Dooge in 1959. The quantitative results were further improved by Mariotte in a publication from 1686. Halley followed Mariotte by contributing to the quantitative analysis of the hydrological cycle. According to Dooge there was no further advance for another hundred years. In Scandinavia, to be exact, it took 84 years. In 1758, Birger Vassenius first made a sound estimate of the components of the water balance and came to correct conclusions about the origin of rivers and lakes. His work, however, is the last in a series of quantitative studies of the water balance of two large Swedish lakes Vänern and Vättern. The first estimates contained several fatal errors that persuaded the authors to keep to the old ideas of water originating from the interior of the Earth. In spite of the progress in quantitative hydrology in the 18th century, the universal recognition of the new ideas in this field, had to wait until the following century in Scandinavia as elsewhere.

1. Introduction

In 1674 a book with the title "De l'origine des fontaines" was published anonymously in Paris. Later it was discovered from the evidence, that the author was Pierre Perrault (Dooge, 1959). Perrault carried out water balance calculations for a basin in the upper part of the Seine and proved by this, that the amount of rain falling in the basin was sufficient to provide for the water in the river. He was the first to test quantitatively, whether the rain falling on the surface of the ground was sufficient to supply water to all the rivers and springs, and thus gave a quantitative argument in the centuries old debate on the origin of springs. Some time later Edmé Mariotte performed similar calculations and published his results in 1686. The flow was measured in the Seine near Pont Royal in Paris, while the rain was measured in Dijon. Perrault's approach may appear rather crude compared to the careful experiments performed by Mariotte. The results of Mariotte were widely circulated and debated in Scandinavia as elsewhere.

Edmond Halley, known mainly for his calculations concerning the comet named after him, carried out detailed evaporation studies. In the publications containing the results of these studies, Halley demonstrated the other part of the hydrological cycle. Among many other things he performed a calculation for the Mediterranean sea and found that the total amount of water brought to it by the rivers, constituted not more than one third of the total amount of water lost by evaporation. He concluded that evaporation from the sea, being a link in a total circulation, is sufficient to supply rivers with water. The works of Halley were also read and debated by Scandinavian scholars.

According to Dooge (1959) there was no further advance in the quantitative analysis of the hydrological cycle for another hundred years and the universal recognition of percolating rain as the origin of springs, was not achieved for a further two hundred years. This description is also applicable to the scientific development in Scandinavia. To be more exact, the first sound estimate of the components of the water balance, that lead to a correct conclusion about the origin of springs, was published 84 years after Perrault by Vassenius (1758). The work by Vassenius is, however, the last in a series of attempts to estimate and explain the water fluctuations of the large Swedish lakes Vänern and Vättern. In addition to these quantitative hydrological studies, the origin of springs, rivers and lakes was a frequent topic of discussion in scientific publications during the 17th and 18th centuries in Scandinavia. The scientific discussions have a local touch reflecting the peculiarities of the Nordic landscape and climate, such as numerous lakes and signs of a higher coastline in the past.

Some aspects of the development of hydrology during the 18th century in Scandinavia have been treated by Otnes (1980, 1981a, b), Gottschalk (1988) and Kajander (1992). Hydrological problems are also considered in the work of Hogbom (1932) and Richter (1959) from a geographical perspective and in those of Frängsmyr (1969), who treats the scientific development of the concepts of geology. Here an attempt is made to present the main contributions to quantitative hydrology in Scandinavia during the 18th century and the reaction of the scientific community to the new concepts about the origin of springs.

2. Hydrometeorological observations

During the 18th century the first regular meteorological observations were started in Scandinavia. Tycho Brahe had been the pioneer in this field. He performed regular weather observations on the Isle of Ven in Öresund during 1582–1597. Tycho Brahe's observations are in the form of thorough weather descriptions in words, as neither thermometers nor barometers were available at that time.

2.1. Observations of temperature and precipitation

In 1664 Rasmus Bartholin was asked to comment on Brahe's weather observations, which had been brought back to Denmark from Johan Kepler's estate in Prag.

He was assisted in this work by Ole Christensen Römer, whose main interest was astronomy but who worked in a broad scientific field. At the beginning of the century he spent much time on different experiments with thermometers. On one occasion, Fahrenheit visited him and took part in these experiments and on returning to his country, started to produce thermometers (Ronan, 1988). The longest continuous weather observations that Römer performed, were for the cold winter of 1708–1709.

In 1751, another astronomer, Peder Horrebow, took the initiative in starting regular observations in the Rundetarn-observatory in Copenhagen. His data are rather reliable, beginning in 1768. One year later daily precipitation observations were also started.

Anders Celsius, also an astronomer, is known as an initiator of permanent weather observations, which included recording of temperature and precipitation among other things. These began at Uppsala in 1739. Separate observations had been carried out since 1722. It is remarkable that, at first, the centigrade scale, named after Celsius, was not used. Several years later, after Celsius's death in 1742, it was finally agreed to use this scale exclusively (Nordenmark, 1936).

Permanent weather observations were also started at this time in other cities of Sweden and Finland, for example, in Turku (Åbo) in 1750. Observations at this latter station were performed by a professor of medicine Johan Leche (1763), who was the first to use rain gauges made of metal instead of wood. Peter Kalm, who continued these observations after Leche's death, wanted to increase the observation network. He considered ordinary wooden buckets to be quite useful in those cases where rain gauges were not available (Kalm and Gudseus, 1754). Observations were more or less continuous at Uppsala, Stockholm, Åbo and Lund, while in other places observation periods were rather short.

A Norwegian teacher Fr. Chr. Holberg Arenz in Bergen was the first to start precipitation observations in Norway in 1765. They were continued for six years. More regular observations were started at the Astronomy Observatory in Oslo in 1839. In Iceland, Niels Harrelow, a solicitor, carried out meteorological observations in Bessastadir near Reykjavik during 1749–1751.

2.2. *Evaporation experiments*

Evaporation was not among those variables that were observed regularly. However, Nils Wallerius performed evaporation studies for five years and published three reports (Wallerius, 1739, 1746, 1747). He experimented with different evaporation pans and he came to the conclusion that for proper observations a cylinder insulated on the sides with clay should be used. The evaporation loss is then directly proportional to the exposed water surface. He also stressed the importance of weighing the cylinders to measure water loss and avoiding the use of a geometric scale which gave uncertain results.

Wallerius does not show any continuous records of daily evaporation except for episodes of a maximum of eight days, for different months. At the same time he observed air temperature, atmospheric pressure and wind. From this he deduced that temperature and wind had a significant influence, while pressure had not. He stressed the importance of evaporation, due to which a considerable amount of water was lost from lakes, rivers, creeks, bogs and the sea, to the atmosphere, and which later returned as precipitation. Wallerius continued his studies by comparing water losses from water, snow and ice. He also mentioned that he had started to study evaporation from soil, but abandoned this for another subject, namely the evaporation from an egg.

3. Water balance studies

Urban Hiärne, who introduced the empirical method in Swedish natural science, was also the initiator of the first quantitative hydrologic studies (Gottschalk, 1988). Urban Hiärne did not make any direct hydrological observations or experiments, but was keen on natural phenomena during his many travels, and he experimented eagerly, especially in chemistry. In a questionnaire (Hiärne, 1694) and when reporting and commenting on the answers (Hiärne, 1702, 1706), he presented in fact the first handbook in geophysics and geography.

The "extremely high levels of lakes" are among the phenomena mentioned by Urban Hiärne. According to the prejudices of the time, this was interpreted as foretelling war or some other great event. While answering the fifth question from the questionnaire, he named as an example, the very high levels of lake Vänern in connection with the spring flood of 1700, when the peasants foretold a great war, which turned out to be correct. Hiärne was, of course, very doubtful about the conclusion and pointed out the necessity of allowing some reasonable people to investigate whether natural causes could be found for the phenomenon. Some 15–20 years later at least three people took up the systematic study of lake fluctuations: Emanuel Swedenborg and Birger Vassenius in Vänern and Daniel Tiselius in Vättern.

Emanuel Swedenborg is the most fascinating personality of all those mentioned here. Before becoming a visionary and prophet, he was actively engaged in scientific work in a broad field of natural history, which had a fundamental importance for his later visionary ideas (Lamm, 1915). His hydrological study is, of course, only a small parenthesis in an attempt to find the grand, world-explaining hypothesis. The results are described in a manuscript "About the fall and rise of lake Vänern and whether this could be related to water recharge and discharge through streams" (Swedenborg, 1907). A dissertation was published on the subject of lake fluctuations in "Acta Literaria Sueciae" 1720–1924, which was probably written by Swedenborg, and where the existing theories on the matter are mentioned (Anonymous, 1720–1924).

Swedenborg took his discharge measurements on the Göta river, at the outlet of Vänern, from the bridge at Rännum not far from the town of Vänersborg. In the manuscript, the width of the river is noted as 14 fathoms (25 m), and the depth varies between 2 and 8 "alns" with an average of 6 "alns" (3.6 m). Swedenborg measured the flow velocity with a chip, noting the time with his pulse, and found it to be 3 "alns" (1.8 m) per second. The total discharge is thus 28 cubic fathoms (158 m^3/s). This is a very low value when compared to our present knowledge. Swedenborg claimed that there were 24 main tributaries to lake Vänern. When estimating the size of the lake, he evidently used old sources, stating that the length was 20 miles (210 km, an old Swedish mile = 36,000 feet) and the width 16 miles (170 km), and thus found the total area to be 320 "kvadrat mil" (36,560 km^2). The real figure is 5600 km^2.

Swedenborg calculated that if the lake's outlet, the Göta River, were closed by a dam, it would take the lake 13.5 years to rise by 1 "aln". From this he concluded that the streams flowing into the lake could not be the main cause of the fluctuations. He then turned to speculations about other causes. A characteristic of the scientist Swedenborg, given by Lamm (1915) is worth mentioning here. Lamm writes: "despite the respect for experiment that he emphasizes, he never has the patience in reality to stop at that. From the very first moment he is thrilled by grand, world-explaining hypotheses".

David Tiselius, a clergyman from Hammar, situated near lake Vättern, is the second person to attempt an estimate of the water balance of a lake. He was given the responsibility for daily observations of the level of lake Vättern by the Scientific Society in Uppsala. Tiselius used a staff gauge divided into 24 inches, mounted on a vertical cliff at the lake shore. He performed his observations between 1720 and 1722. According to Wallen (1911), these are the oldest preserved observations of lake level in Sweden. Tiselius worked intensively to clarify the fluctuations of lake Vättern, which is evident from the records of the Society. (Hag, 1983, gives a detailed description of these studies). In March 1723 the manuscript of his report "A complete description of the large Swea and Giotha lake Watter, its position, size and remarkable properties" was published (Tiselius, 1723). His conclusions were seriously criticized by the Society, not in respect of his observations, but of the more speculative parts.

In 1730 a second report was published "Further experiments and lake measurements in Wätter, on the occasion of remarks and remembrances of learned men" (Tiselius, 1730). The reports contain a very thorough description of the lake's geometry, rejecting the rumours about Vättern being bottomless. Furthermore, they contain a water balance calculation, inspired by Swedenborg's work. The flow velocity was measured by allowing a boat to float a distance of 600 feet along the Motala river at the outlet of the lake. From this Tiselius calculated the discharge to be 59 m^3/s, which is an acceptable figure from our present knowledge. Tiselius used this single observation to estimate the time necessary to lower the lake level by 2 feet — 2 years and 9 days, in the case of no inflow. From this Tiselius concluded that so much water could not by any means come into the lake from the 40 minor streams, that he had noticed as tributaries. This made him suggest that there existed subsurface channels that contributed to the main part of the inflow. In a similar way to Swedenborg, mistakes in the water balance calculation lead Tiselius away from the proper answer.

Birger Vassenius is the third to take up the challenge of explaining fluctuations in water level. Together with his brother he carried out meteorological and water level observations of lake Vänern from 1714. He produced a record of observations from 1698, by including older information. These observations were published for the first time in a dissertation "Cagitationes de incremento et decremento lacus Vener" (Vassenius, 1730).

The published record describes only qualitatively the moisture conditions during different years. The conclusions drawn from these observations were translated into Swedish by Wallen (1911). According to these, the local conditions around Vänern, for instance, dry summers or springs with intensive snow-melt, do not influence the fluctuations. Neither did he find that the flow at the outlet through the Göta River could explain the variation. All his findings are correct, but when he turned to explanations, he kept to the traditional theories supported by his learned colleagues.

However, Vassenius continued his observations and in 1758 he published a second article about Vanern "Remarks on lake Vänern" (Vassenius, 1758). He had not totally given up his earlier beliefs and explanations. He could not deny subsurface channels as they had been witnessed, according to him, in the Holy Bible. However, he came to the conclusion that they could be doubted in the case of Vänern. He claimed therefore that the variations in the level of the lake were explained purely by meteorological conditions, an explanation that agrees with our present concepts. He commented on the influence of rain and snow-melt and mentioned especially that it was not only the local weather conditions around Vänern that were of importance, but also those in the Norwegian mountains, more than 500 km to the north.

Vassenius was also interested in evaporation problems. Based on their own meteorological observations, the Vassenius brothers related the evaporation losses to weather conditions. Birger Vassenius claimed that on a nice summer's day the evaporation could be as much as 1/2 inch (1.2 cm). He then made the same assumption as Swedenborg, that the momentary inflow from tributaries equalled the outflow to the Göta river, to show the importance of evaporation. The water level would be lowered by 1 aln in 40 days under this simplifying assumption. Vassenius changed his views about the causes of the regular fluctuations of lake Vänern, after he collected another 28 years of observations. An important issue was to understand the influence of evaporation from the lake.

4. The origin of springs

Before the change from speculative science to a more empirical approach, hydrological subjects played a significant role. The problem of the origin of springs and rivers is perhaps one of the most frequent topics in Scandinavian university dissertations during the 17th century. According to Richter (1959), the first work originates from 1618 and was followed by at least another ten on the same topic, from the universities of Uppsala, Lund and Åbo. Frängsmyr (1969) characterizes these works as "Spiritless retelling of foreign authors, names like Varenius and Kircher mentioned ad nauseam".

Scientific development in Scandinavia was closely connected to the events in the rest of Europe. Most scientists mentioned here spent many years studying and working in other European countries, especially France, Germany, the Netherlands and England. Several of them were members of foreign scientific societies. Foreign scientific works by authors like Athanasius Kircher, Bernhardus Varenius, Edmé Mariotte, Peter von Musschenbroek, Edmond Halley, Robert Plot and Domenico Gugliemini, are frequently quoted in Scandinavian scientific discussions. Mariotte's and Halley's works were thus well known to the scientific community of the Nordic countries, but their pioneering ideas made their way through the barriers of old views very slowly.

4.1. Rain or "underground channels"?

Caspar Bartholin, junior, was the first Scandinavian to support the new ideas on the origin of springs (Kajander, 1992). In two of his works: the dissertation "De fontium fluviorum — que origine ex pluvius" and the text-book "Specimen philosophiae naturalis", he discusses all other theories on the subject and demonstrates that they are impossible. Bartholin spent a great deal of energy to prove that there were enough fissures and depressions in the ground, where water could accumulate, but at the same time that there existed impervious sediment layers along which water could flow to appear later in the light of day. Finally, he used the results obtained by Mariotte, to prove that rain water is sufficient to maintain river flow. Bartholin, born into a family with traditions in science, was a highly respected scientist and teacher. His authority most probably influenced the majority of his students to support the new ideas about the origin of springs.

In Denmark the victory of the new ideas was not complete at this time, and in Sweden their progress was very slow. Urban Hiärne introduced new scientific trends and discussed them thoroughly in his text-books on geophysics and geography. He relied also on the information obtained from his questionnaire and took personal initiatives in quantitative studies. From this background, he formed his own opinion about the physics of the Earth, and his influence on the thoughts of the new generation of scholars was significant. Concerning the hydrological cycle, he repeated the arguments from Kircher to Varenius, as reflected in the Swedish dissertations on this subject from the 17th century. On the other hand, he also presented the new ideas by Mariotte and Halley.

Hiärne (1702) quoted Kircher and Mariotte and discussed the new ideas. He agreed that to discover the real origin of springs was a very delicate matter. He found the existence of the underground hydrophylacia, suggested by Kircher, to be fascinating and in accordance with his own and others' observations of lake Vättern. Important support for this, according to him, could also be found in the Holy Bible. Hiärne finally rejected Mariotte, presenting the following arguments:

– Over a larger area, the rainfall cannot be as uniform as revealed by a single rain gauge, but has significant variability.

– If the river flow were to have its origin in rainfall, it would not vary as slowly as it does in nature.

– It is unbelievable that evaporation could be as much as one third of the precipitation.

– There are very large fluctuations in rainfall and river flow between different years, making Mariotte's findings unrepresentative.

Hiärne's arguments reveal a good understanding of hydrological processes with the exception of evaporation. His misunderstanding of the latter led him to make erroneous conclusions.

Pehr Elvius (1748), in a dissertation "Water evaporation" gave a very complete review of contemporary works on the subject, quoting Mariotte, Halley, Sedileau, Musschenbroek and others. It seems that the experiments by Nils Wallerius for Nordic conditions greatly influenced him when he admitted the importance and magnitude of evaporation in the hydrological cycle.

Carl von Linné also turned to the problem of the water balance in his numerous scientific works. In his "Oeconomia Naturae" (Linné, 1906a), published in Swedish in 1750, he gave his view on the origin of the components of the water balance, though without reference to hydrological findings. He excluded at once speculations on subsurface processes, by saying "One who wants to amuse himself with guessing games, can find them with Cartesium, Kircherum, Helmontium". Rainfall obtains its contribution from evaporating vapours that are carried around in clouds, until they unite into drops and are brought to the earth. The origin of springs is rain water that percolates through the soil. Rivers in their turn start from springs. They bring the water back to the sea, from where it initially rose, so that it will not be emptied but continuously, as in a ring, follows the same path. Both Pehr Elvius and Carl von Linné, thus, revealed very modern concepts of the hydrological cycle, but their opinion does not seem to have been generally accepted.

Cartesianism was very strong in Sweden at this time and especially strong in the University of Uppsala. This is perhaps the main reason why the new ideas did not make a break-through in this country during the first half of the 18th century. It became, however, more or less accepted in Finland, where Russian occupation hindered university activities during two centuries and effectively eliminated the old cartesians. In 1732 a professor of mathematics, Nils Hasselblom (Hasselblom and Roering, 1732), published the first dissertation on the origin of springs supporting the new ideas, though the scientific struggle was far from over. The old theories found new supporters, and in 1786 battles still took place in the Academy of Åbo. However, in a dissertation from this year on the reasons for floods, the results of Mariotte were quoted and accepted (Gadd and Kekonious, 1786).

Three text-books on hydrology were published in Scandinavia during the 18th century. The first one "Hydrologia eller Watturiket" ("Hydrology or the Water Kingdom") was written by Johan Gottschalk Wallerius in 1748 and forms the basis for the two others: "Naturligt Vande" ("Natural Water") by Johan Chr.

Lange (1756) and "Hydrologia Norvegica eller Norsk Vand-Rige" ("Norwegian Hydrology or Norwegian Water Kingdom") by Christopher Hammer (1778). Both authors, especially Hammer, often refer to Wallerius. The books do not contain much of what we expect to be present in a text-book on hydrology. They describe mainly classification systems for the liquids, everything from resin to wine, and are remarkably influenced by Carl von Linné's sexual classification system for plants. No references to Mariotte or Halley can be found in them.

Even at the end of the 18th century the new ideas on the hydrological cycle were not generally accepted in Scandinavia. There were still many supporters of the concept of "underground channels". An Icelander, Eggert Olafsson, explained, for example, the growth of glaciers by means of underground channels to the sea (Thorarvinson, 1960).

4.2. "Disappearance of water"

To understand why it was so difficult for the new ideas on the origin of springs to become accepted in Scandinavia, it is necessary to mention a special subject that dominated scientific discussions at that time, and which can perhaps, explain why the old-fashioned ideas survived so long. The question of "Wattuminskningen" ("Disappearance of water") occupied the minds of the Nordic scientists. Naturalists discovered obvious evidence that the land surface had been submerged under water in the past. Landmarks, known to have been situated on the shoreline, were now found far inland and remains of beach-ridges and shell banks could be observed at altitudes up to 200 meters above sea level. Everybody considered it to be the evidence of the deluge, but nobody could explain why the water disappeared subsequently. The idea of land elevation as a probable explanation had still to come. Ephraim Otto Runeberg (1765) from Vasa wrote about land uplift in 1765, but was completely ignored. More generally, this question was discussed in the 1720s.

The "Disappearance of water", after it had been first mentioned by Urban Hiärne, was also treated in a paper by Emanuel Swedenborg (1718). Twenty four years later, the subject was taken up again by Anders Celsius (1743) and in the same year also by Carl von Linné, who discussed this problem in a speech (Linné, 1906b). It was, however, not until the historian Olof Dalin (1747) used the theories in his work on the history of Sweden "from outset to our times", that the controversy started, mainly because they were not in agreement with the Holy Bible. The controversies engaged many scientists and clergymen. In a biography from those days (Warenholtz, 1783) thirty-three books and papers are named that directly treated this subject. As such, it is not directly related to hydrology, but the different theories constructed to explain the phenomena influenced concepts about the hydrological cycle. Many works on natural history, including hydrology, were written, and the views of the controversy on the disappearing water can be traced in all of them.

The old ideas gave a possible explanation for the "disappearance" of water by "underground channels" that could have transported it away, or it could have been transformed into soil or some other matter. The new theory could not give any explanation. "Underground channels" and "hydrophylacies" are still discussed in physiographic publications in the middle of the 18th century. For example, in a text-book from 1883 (Klein, 1883) there is a presentation of these theories, though they are then called "old-fashioned".

In the dissertation "Remarks on the diminishing of water", by Anders Celsius, two causes are given: water remains in vegetation and turns into soil as it decays and, in accordance with Urban Hiärne, there are holes in the bottom of the sea and water fills up the interior of the earth through them. Emanuel Swedenborg (1718), Carl von Linné (1906b) and others, were all convinced of the disappearance of water, but they gave no explanation and mainly collected and described the evidence for disappearing water. It was Johan Gottschalk Wallerius who at the end of his life gave the most thorough treatment of the problem in the book "Thoughts about the world and especially the creation of the Earth and change", published in 1776. At this time, however, his ideas were obsolete. He neglected totally all newer hydrological works, Swedish as well as international, and his main goal was to combine the biblical creation with his scientific ideas about water. He described five different experiments to obtain soil from water, but as a true neptunist he also claimed that water could be turned into air, salt, oil and sulphur.

The contemporary antipode to Johan Gottschalk Wallerius is Torben Bergman, a very open-minded, realistic and modern scientist. In his "Physical description of the globe of the Earth" (the second edition was published in 1774), the subject "water" is treated most thoroughly as well as "water disappearance". When discussing the origin of springs, he incorporated all the different opinions. He hesitated about which ideas to support and chose a middle road of compromise, sometimes accepting subsurface channels as the cause for water circulation, and at other times giving preference to precipitation and evaporation. His attitude is similar towards the problem of disappearing water. He found the evidence convincing that the land surface had been covered with water, but when looking for causes, he left the question open for future solution.

5. Conclusions

During the 18th century several observations and experiments were made in Scandinavia to quantify hydrological processes. Due to unfortunate circumstances mistakes were made in attempts at water balance calculations. The most serious one was neglecting the importance of evaporation, although very serious studies on this matter had already been carried out. An analysis of these mistakes is important for understanding why the modern concept of the hydrological cycle was not generally accepted by scientists. The theories explaining the "water disappearance" also preserved the traditional 17th century concepts on the origin of springs.

The practical use of research results is an important factor for scientific development in 18th century Scandinavia. Carl von Linné had undertaken his scientific travels to describe natural resources in order to use them better. Numerous pupils and successors followed his example. In many respects, however, the new knowledge about nature remained the privilege of a limited circle of educated people. Theory and practice followed different paths of development. Nearly 150 years passed before hydrology reappeared as a scientific discipline at Scandinavian universities.

Historically, technique has always been regarded as handcraft. Technical schools, as a rule, emerged from handcraft schools in the beginning of the 19th century. Even today we are inclined to regard technique as applied science, though it has its own development. It is obvious that great achievements in hydrology were made possible as a part of technical progress, when there was a need for the technical solution of particular problems: construction of channels, reclamation of lands by lowering the level of lakes, provision of adequate water supply, construction of dams and hydropower plants etc. This development became more intense at the beginning of the 19th century, but had old traditions. It was independent of the debate of learned men on the origin of springs. A concrete example of this new development is the initiation of one of the longest series in the world of continuous hydrological observation. In 1805, Baltazar von Platen, the principal contractor of the Göta channel from Stockholm to Gothenburg, began daily observations on water levels in Vänern at Frugärden (two years later moved to Sjotorp) (Wallen, 1911). The observations still continue, but were disturbed by flow regulation from 1931 onwards (Melin, 1970).

References

Anonymous, 1720–1724. De celebribus utriusque Gothiae lacubus, Vennero Vettero gue Observata guidam ab crudutus viris per literas transmissa etc. Acta Literaria Sveciae, 1: 116.

Bergman, T., 1774. Physisk Beskrifning öfver Jord-klotet. Del I och II, Andra upplagan. Tilökt och förbätrad, Upsala.

Celsius, A., 1743. Anmarkningar om vattnets forminskande så Östersjön som i Vesterhafvet. K. Vetenskapsakad. Handl.

Dalin, O., 1747. Svea Rikes historia. Del 1, Stockholm.

Dooge, J.C.I., 1959. Un bilan hydrologique au XVIIe siècle. La Houille Blanche, 6: 799–807.

Elvius, P., 1748. Vetenskapernas historia, Del 1. Stockholm.

Frängsmyr, T., 1969. Geologi och skapelsetro. Lychnos-Bibliotek 26, Stockholm.

Gadd, P.A. and Kekonius, N.J., 1786. Undersökning om orsakerna till flodvatnets öfversvämningar i Finland. Åbo.

Gottschalk, L., 1988. The dawn of quantitative hydrology in Sweden. Nord. Hydrol., 19: 165–182.

Hag, T., 1983. Vatten, Vättern och Vetenskapssocieten. Föreningen för Västgötalitteratur, Vara.

Hammer, C., 1778. Hydrologia Norvegica eller Norsk Vand-Rige. København.

Hasselblom, N. and Roering, A., 1732. Dissertatio gradualis de origine fontium. Aboa.

Hiärne, U., 1694. Een kort anledning till åtskillige malm- och bergarters, mineraliers, waxters, och jordeslags, sampt flera sållsamme tings efterspöriande och angifwande. Stockholm.

Hiärne, U., 1702. Den korta anledningen, til atskillige malm- och bergarters, mineraliers och jordèslags etc. eftersörjande och angifwande, beswarad och förklarad. Första flock om watn, Stockholm.

Hiärne, U., 1706. Den beswarade och förklarade anledningens andra flock, om jordens och landskap i gemen. Stockholm.

Hogbom, A.G., 1932. Naturbetraktelser i historiskt perspektiv. Uppsala.

Kajander, J., 1992. The acceptance of pluvialism in the Nordic universities. Proc. XVIIth Nord. Hydrol. Conf., Alta. (In press).

Kalm, P. and Gudseus, A., 1754. Oforgripeliga tanckar om sattet att anstalla meteorologiska observationer och theras nytta i ecconomien. Åbo.

Klein, H., 1883. Jorden og dens naturforhold. Kjøbenhavn.

Lamm, M., 1915. Swedenborg. En studie öfver hans utveckling till mystiker och andeskådare. Stockholm.

Lange, J.C., 1756. Naturligt Vande. Kjøbenhavn.

Leche, J., 1763. K. Sven. Vetenskapsakad. Handl., pp. 15–17.

Linné, C. von, 1906a. Skrifter af Carl von Linné utgifna af Kungliga Svenska Vetenskapsakademien II, Valda smärre skrifter af allmant naturvetenskapligt innehåll. Stockholm.

Linné, C. von, 1906b. Om den beboeliga jordens tillväxt. Tal hållet vid medicine doktorspromovationen den 12 april 1743. Övers. T.M. Fries i Skrifter, 2, Uppsala.

Melin, R., 1970. Hydrologi i Norden. Utbildningsförlagèt. Stockholm.

Nordenmark, N.V.E., 1936. Anders Celsius. Lychnos-Bibliotek I, Uppsala.

Otnes, J., 1980. Nordens eldste hydrologibøker. In: Nordisk hydrologi i utveckling. SMHI, HB-Rap. Nr. 42, Norrköping.

Otnes, J., 1981a. Where had all the water gone?. Geophysica, 17: 1–2.

Otnes, J., 1981b Swedish influence on Norwegian hydrology prior to 1814. Geogr. Ann., 63A(3–4): 119–125.

Richter, H., 1959. Geografiens historia i Sverige intill ar 1800. Lychnos-Bibliotek 17(1), Uppsala.

Ronan, C., 1988. Histoire mondiale des sciences. Éditions du Seuil, Paris.

Runeberg, E.O., 1765. K. Sven. Vetenskapsakad. Handl., pp. 81–115.

Swedenborg, E., 1718. Om watnets hogd och förra werldens starcka ebb och flod. Bewjs utur Sverige. Stockholm.

Swedenborg, E., 1907. Om Wennerns fallande och stigande och hurwida det härröra kan af wattnets tillopp eller aflopp genom strömmar. In: Swedenborg, Opera guaedam aut inedita aut obsoleta de rebus naturalibus 1. Geologica et epistolae. Stockholm, pp. 33–40.

Thorarvinson, S., 1960. Glaciological knowledge in Iceland before 1800. Jökull, Reykjavik.

Tiselius, D., 1723. Uthförlig beskrifning öfver den stora Swea och Giotha siön wätter, til dess belägenhet, storlek och märkvärdiga egenskaper samt anmärkningar och berättelser om några uthi och omkring sion belägna öjar, näas, berg och strömar; med mera, som wärdt är att i ackt taga. Efter åhtskilliga förnäma wänners åstundan sammanfattad. Upsala.

Tiselius, D., 1730. Ytterliggare försök och siö-profwer uthi Wättern i anledning af lärde mans anmärkningar och påminelser, wysande wattnets nogare beskaffenhet, af de grundsatser som i Watters beskrifvningar aro anforde. Stockholm.

Vassenius, B., 1730. Cogitationes de incremento et decremento lacus Vener. Acta Literaria et Scientiarium Sueciae, 3: 16–22.

Vassenius, B., 1758. Inmärkningar om insjön Wänern. Den Swenska Mercurius (bihang), Arg. 4(1): 406–415.

Wallen, A., 1911. Kännendomen om Sveriges hydrografi före tiden omkring 1870. Hydrografiska Byrån, Årsbok 1908–09, Stockholm.

Wallerius, J.G., 1748. Hydrologia eller watturiket, indelt och besrifvit, jaäte anledning til wattuprofvers anställande. Stockholm.

Wallerius, J.G. Tankar om Verdens I synnerhet Jordens Danande och Åndring. Stockholm.

Wallerius, N., 1739. Försök om dunster uppstigande uti lufttoma rum giorde och till academien insände. K. Wetenskapsakad. Handl., pp. 282–292.

Wallerius, N., 1746. Rön, hvarigenom åtskilliga Naturens lagar, angående Vattnets och flytande Materiers utdunstandet frambringas. K. Wetenskapsakad. Handl., pp. 1–21.
Wallerius, N., 1947. Rön om isens utdunstande, utgiörandes den tredie Afhandlingen ut i detta ämne. K. Wetenskapsakad. Handl., pp. 214–226.
Warenholtz, C.G., 1783. Bibliotheca Historica Sveo-Gothica. Stockholm.

Chapter 12

Hydrology and hydrologists — reflections

J.E. NASH

Abstract

Hydrology is presented as a science, a domain of application of science and a technology. It is advanced by the application of known principles of more basic sciences but also by empirical analysis. The paper speculates on the future role of hydrology and hydrologists and the importance of environmental awareness.

1. Introduction

When in the middle 1950's I first began to describe myself as a hydrologist the term was not well understood generally and the category was sparsely occupied as few engineers or other professionals would accept the designation because of the restrictions which it implied. It is true, that since the 1920's there had been a branch of scientific hydrology within the International Union of Geodesy and Geophysics and since 1930 the American Geophysical Union had accepted hydrology as its seventh section. These designations did not however, or at least had not by the middle fifties, created an acceptance or awareness of the profession of hydrology. Indeed the very use of the adjective *scientific* in the title would seem to have been defensive, or to constitute a claim which the other branches of the unions did not feel it necessary to make.

Over the intervening years this situation has changed greatly. There are now, in most countries national hydrological organisations and many professional persons are happy to describe themselves as hydrologists. Most engineers have come to recognise engineering hydrology as an area of specialisation within the broad engineering field rather than merely a technical activity. Now indeed many writers, including the distinguished former president of the International Association of Scientific Hydrology, Dr. V. Klemes, decry what they see as an engineering takeover of what should be a geoscience. In 1989 at the instigation of Dr. Klemes, the IAHS and UNESCO jointly appointed a panel to consider

> "whether the education of hydrologists is suffering from a lack of recognition of the natural science base of hydrology and, if deemed necessary, to make proposals for new emphases in the organisation, structure and content of

University education of hydrologists, within the ambit of geoscience and of applied professional practice."

I was chairman of that panel and I know how hard my distinguished colleagues worked to reach and express considered and useful conclusions. Compromise was necessary. We were never unanimous, except in our willingness to compromise, and we never expected to write "the law of the Medes and the Persians which altereth not". I am sure that my panel colleagues will not resent it if I now express my own views, conditioned and informed indeed by their deliberations and perhaps matured by the passage of a few years since our discussions.

2. The alleged failure of engineering hydrology

The report of the IAHS-UNESCO panel (Nash et al., 1990) reflects the widely held view that hydrology as an engineering specialisation has at least partially failed and that future progress must depend on recognition of hydrology as a geoscience to be developed, practised and taught in geoscience departments of universities. To me the most significant part of this report is the explicit claim to a scientific as distinct from a technological status for hydrology, implying a second step from the pre-1950 situation and perhaps fulfilling the aspirations of the founders of the International Association of Scientific Hydrology.

Yet hydrological activities necessarily involve purely technical components, such as flow measurement, technological development involving the provision of answers or means of obtaining answers to real life problems in water resources development, as well as science. This is not an uncommon situation. A corresponding division obviously exists in medicine and probably in many other professions. Why then do hydrologists agonise over the status of their subject and seek, as they appear to have done for the last seventy years at least, the status implied by the term *scientific*?

It might indeed be said that the IAHS-UNESCO report seems to attribute all the ills of the present state of hydrology to the engineering emphasis of its foundation, and to rely excessively on the naive expectation that all will be well when responsibility for training and education in hydrology is transferred from engineering to geoscience departments of universities.

While most of the current criticism of the state of hydrology and hydrological education originates within the family, we are also assailed from without, by environmental lobbyists, for alleged narrowness of outlook and for our inability, or at least our failure, to assess the consequences of our interferences in the watery environment. To meet the criticism from within and without we must consider and decide for ourselves, what hydrology is, where it lies in the spectrum of human knowledge and endeavour, between the extremes of mere technical application and the deeper understanding of science. Without this concept, we cannot begin to consider systematically what a hydrologist is, or what would be an appropriate system of education and training for students of hydrology.

3. Science or technology

Any science is a body of related knowledge, pursued and acquired in a generally agreed way — the scientific method. In some domains, the term science is applied to mere classification of species or observations, or even to a description of what is in the domain. In others — the "harder" as distinct from the "softer" sciences — the emphasis is on analysis and deduction, and reason rather than observation, is the primary faculty used in expanding the body of knowledge. The term science is however also applied to the activity of obtaining the knowledge as well as to the body of knowledge itself. In this sense, science is an activity of the intellect rather than a mere compilation of knowledge in a prescribed domain. Technology, on the other hand, refers to doing rather than understanding; but the development of a technology may be as scientific as the advancement of a basic science. The intellectual activity is not dissimilar even though the end is different. I think it is much more important for us to ensure that the activity by which our knowledge of hydrology is advanced is scientific, rather than to insist that the knowledge obtained implies an increase in understanding.

Hydrology as a science might be defined as knowledge of the occurrence, state and behaviour, particularly the movement, of water in the natural environment. If this knowledge is restricted to mere description of where and how water occurs and moves, hydrology would be a soft science, a matter of classification, advanced by mere observation, and while this species of knowledge might well be useful, it would not be very interesting. On the other hand, if the knowledge referred to implies an understanding of the laws of nature relevant to the watery domain, hydrology as a basic science would seem to have little room for scientific activity, at least at the particle level, since these laws are already well understood.

Hydrology as an applied science. Not being a basic science, and not wishing to be classified as a mere categorisation, hydrology if scientific, would seem to be so because it provides a domain of application wherein basic scientific knowledge is applied to investigate, explain or predict some phenomena. One might however question whether such an application should be called hydrology or be known rather by the name of the basic science which is applied. The application of the principles of physics to produce knowledge in the domain of water might give rise to a body of knowledge called hydrology, but the activity would be physics. The fitting of frequency distributions to flood magnitudes is a statistical not a hydrological exercise.

If only one basic science were involved, the term hydrology might be unnecessary, except to designate the compiled knowledge or the domain of application of the basic science. Because more than one basic science is involved hydrology as an applied science acquires many facets and the role of the hydrologist is widened. The solution of operational problems may require the application of many different sciences, but invariably, it also requires the prior application of the imagination in the recognition of the problem and its formulation in a quantitative manner and subsequently in the creation of hypotheses or models which involve the synthesis of mathematical structures which reflect the interaction between the variables identified. This is a scientific activity in the domain of hydrology which neither the mathematician, the statistician or the physicist inexperienced in the solution of hydrological problems, can undertake though each can contribute to the solution once the problem has been identified and quantified by the hydrologist.

This is a very important role, akin to that of the conductor of an orchestra who assembles the different instrumentalists, defines their activities without usurping them, controls their interaction and ensures an harmonious whole. There is however another, more obviously scientific, role for the hydrologist which tends to give hydrology a special status.

4. Empirical hydrology

At the catchment and higher scales, the complexity of the boundary conditions often renders it impracticable to apply even well known natural principles to obtain solutions to real life problems. The obvious expedient of taking spatial averages is not always available either. Relationships between averages are not necessarily the same as those between point values, as consideration of the relationship between amounts of rainfall, runoff and infiltration immediately shows. There is thus a legitimate role, for the hydrologist, to seek by observation and analyses of records, empirical relationships at the catchment and higher scales, rather than to rely on the synthesis of these from basic physical laws. The method of this empirical, parametric, engineering or "real" hydrology is that of a researcher in extending a basic science, where progress is made by hypothesis formulation, testing and modification, within the appropriate domain, rather than by deduction from a more basic science. Thus, the hydrologist acquires a research role differing from that of the basic scientist only in the level of application, viz., the catchment rather than the particle.

5. The two pillars of hydrology

I suggest that it is this two-fold role, this dependence on two pillars, which distinguishes the work of the hydrologist from that of the physicist, chemist or botanist engaged in the application of his science in the domain of water. Both pillars are necessary. Exclusive reliance on either is unnecessarily restrictive.

It is probable that, in the past, engineering hydrologists have tended to rely excessively on the pillar of empirical analysis and it is fashionable to attribute the limited progress which has been made in hydrology over the past forty years to this defect. I do not fully subscribe to this view, nor do I accept the naive assumption that as soon as we recognise hydrology as a component in the earth sciences, and treat it accordingly, the solutions of the real life problems of hydrology will become manifest. It may be as false in the hydrological as it has proven to be in the economic context, to assume that the rising tide must raise all boats. Even if, however, the assumption is valid, is it appropriate to blame the engineering hydrologists for their failure to be other than engineering hydrologists? Does not the blame, if blame there is, lie with the geophysicists who failed to apply their superior scientific training and experience in the hydrological context?

I think hydrologists, generally, have shown a remarkable willingness to learn new skills and to borrow tools of analysis developed in other contexts. In the last hundred years hydraulic engineers have shown a remarkable facility in applying the laws of fluid mechanics where these were available and a reliance upon the empirical method, when they were not. Until approximately one hundred years ago, the importance of the conduit roughness in the resistance to turbulent flow was almost completely ignored but when the importance of roughness was recognised, the simple empirical relationship known as the Chezy-Manning equation, which is adequate for all practical purposes, was quickly found. Subsequently, boundary layer theory leading to the logarithmic velocity distribution and the experiments of Nikuradse, produced the "rough and smooth pipe laws" and the Colebrook-White equation — a remarkable development, intellectually satisfying, broader in its application than the Chezy-Manning formula, but not really more useful in the domain of fully developed turbulence. If they had been differently trained, Manning and Chezy might not have discovered what we now call the Colebrook-White equation, but they certainly would not have simply waited for the development of fluid mechanics to provide a good physical basis for their work. The lesson, of course, is that we must depend on our own resources and make such progress as we can in the prevailing circumstances of knowledge.

6. The real failure of engineering hydrology

We may have neglected the possibility of developing hydrology as an application of more basic sciences. This however does not imply that there is something wrong in seeking the macro relationships of hydrology through the analysis of observations and records — a method which might seem more appropriate to the advancement of a basic science. It is not for this that we must be blamed, but for our lack of scientific discipline in the application of the scientific method and for indulging a fascination with the tools of analysis themselves, rather than concentrating on their use. We have tended to be careless in the measurement of our basic quantities — the accuracy of rainfall measurements is still suspect and we have failed to obtain a method for the direct measurement of evaporation as a vertical vapour flux. We have been content to use the concept of the catchment as a control volume, within which the mass and energy fluxes are assumed to be conservative, but we have often neglected the real possibility of water leakage between catchments and the advection of energy.

We have not always tested our hypotheses of catchment relationships, nor do we normally examine the errors resulting from their application, to obtain indications for further modification. On the contrary, we have tended to settle for any hypothesis which provides a fair reproduction of observed behaviour. On occasion we have departed from all consideration of observations and concentrated on the refinement of the tools themselves, thus creating the subject of "blackboard hydrology" — a field of study which though providing many opportunities for superficial brilliance, has little or no prospect of ever becoming relevant to actual hydrology. This subject remains with us today. A most recent manifestation has been the tendency to express catchment behaviour in probabilistic terms even when no probabilistic concept is involved. The instantaneous unit hydrograph has become "the probability density function of the waiting time". Blackboard hydrology proceeds by the adoption of new clothes rather than more solid achievement, and, by and large, the participants at international conferences and the editors of journals are reluctant to assume the responsibility of decrying the practice.

7. The future for hydrology

Reflecting the recognition of hydrology as an interdisciplinary subject, there has been, in recent years, an increasing reliance not on hydrologists trained as such, but on teams of scientists trained in more basic disciplines. Major problems of research and application in hydrology are increasingly being entrusted to such teams. This development is most natural and will undoubtedly be the method by which hydrology is advanced and the major problems of application tackled in the future. Within such teams there remains, of course, the coordinating role for the hydrologist, for which his experience should render him particularly fit. The importance of his second, empirical, analytical, role however may diminish, the further the problem being tackled is beyond existing experience. Hydrologists will have to learn to work with and coordinate the work of interdisciplinary teams which will be assembled for specific purposes or for general hydrological research. The alternative of creating a Renaissance type hydrologist, competent in all the basic sciences and experienced in their application, is to ask too much, at least for the immediate future. If hydrologists fail to meet this challenge, however, they may find that there is no place for them on the interdisciplinary teams and the profession of hydrologist may well disappear.

8. The education and training of hydrologists

Perhaps unique among the practitioners of applied sciences, hydrologists have not generally enjoyed a specific, well-defined, training. It is probably true that most of the pioneers of hydrology, particularly those distinguished hydrologists of the immediate post war period, had little undergraduate training in hydrology and probably very little formal training in hydrology at any level. The ranks of hydrologists have always been replenished by a drifting in from other disciplines, and it must be recognised that many who came to hydrology in this way brought with them a consuming interest, a lively intelligence and sometimes an acquaintance with technologies which were relevant but previously unused in hydrology.

This however is not an orthodox way in which to build a profession. It would be better for hydrology if its distinct status were recognised and its practitioners trained in an appropriate manner, though not necessarily at the undergraduate level. The employment opportunities for hydrologists may be too few to justify such an early career commitment. However, there is no reason why graduates in the physical sciences generally (in various mixes) should not become hydrologists through subsequent specialisation by way of postgraduate courses. Astronomers are so trained; so generally speaking are geophysicists, using the term in the narrower sense. Geologists, however, are usually required to make a commitment at the undergraduate level. Perhaps because their subject is relatively descriptive, a less intensive training in the basic sciences is sufficient, thus leaving room for the inclusion of some geology in undergraduate curricula. Hydrologists, however, need to learn a wide range of basic scientific material, become familiar with the scientific method and acquire the technological expertise of a profession. It is perhaps not surprising that the somewhat haphazard educational system which has evolved rather than been designed, has partially failed to meet these requirements. The IASH-UNESCO panel to which I referred, distinguished between scientific hydrology and scientists on the one hand, recommending that students should be trained in university departments of geoscience, and applied hydrology and professional hydrologists on the other, recommending that

> "they should continue to be trained in postgraduate courses which should however provide an increased content of basic sciences and a greater variety of hydrological sub-fields. ... efforts should be made to ensure that the developments in hydrological science achieved in geoscience departments are made available to the applied hydrologist, typically in engineering departments. Close contact should be established (between the two training systems)."

I confess that I would now prefer a single hydrological stream (only exceptionally at the undergraduate level) normally drawing from graduates in engineering and science and providing for longer and more scientific training in the relevant basic sciences, but recognising also the specific science of hydrology referred to above as the second pillar. I think that the defect in our present postgraduate courses lies largely in the inadequate preparation of the entrants and the inadequate duration of the courses themselves. If we could rely on obtaining good honours graduates in any of many combinations of science subjects or in engineering, and if we had these, not for a single year or a year and a half, but for two or three years, I believe we could overcome all the present difficulties and, in time, the standards would rise so that the graduates would be capable of fulfilling both the scientific and professional roles and the continuation of both through teaching.

Where an undergraduate hydrological training is feasible, I would consider that this might be provided as an option within a broader civil or environmental engineering framework. With a greater emphasis on physics and chemistry, some introduction to plant science, about the present amount of geology, some geophysics and hydrology in the narrow (or basic) sense, a very good training could be provided in a typical four years degree programme if room were found in the syllabus by renouncing the structural engineering elements and other, often optional, courses which are provided in most contemporary civil engineering programmes.

Engineers, trained to this level in their undergraduate courses, would be eminently suitable for postgraduate training as hydrologists, even in relatively short courses. Moreover, they would not lose their designation as civil or environmental engineers, thus preserving their wider career opportunities. I feel that such graduates, within the general civil or environmental engineering fields, would be at least as well trained for employment in those fields as are the present graduates, typically trained on a surfeit of structural mechanics to a level which very few ever use.

9. Environmental awareness

I would like to finish with some reference to the necessity to inculcate an awareness of the environment during the education and training of hydrologists. The work of the hydrologist often severely impinges on the environment. There may be consequences outside the immediate concern or interest of the hydrologist, which on that account may be neglected. That is not to say that there is another science — environmental science — whose practitioners must be given a veto on water resources development or indeed on any development. An adequate training for a general purpose environmental scientist would require a very long period indeed and students of wide interests and abilities. If it is difficult to train hydrologists, it would be even more difficult to train environmental scientists to any useful level. Yet we do need an environmental awareness, not only in hydrology but in many activities. In my view this must be supplied by extending the concern of the specific professional rather than by creating a new profession.

When, a few years ago, the directors of some UNESCO sponsored international postgraduate hydrology courses were asked by UNEP to consider, as an experiment, the inclusion of an environmental component in their courses, it was suggested that this should be done through: (1) the inclusion of a specific environmental impact course, and (2) emphasising environmental aspects in all of the courses taught.

I have no quarrel with the second suggestion, except to say that not all subjects could be taught in a manner involving a consciousness of environmental effect. I feel however, that it is not possible to define a suitable course to meet the first suggestion. The requirements of such a course would be so wide, and to be useful so deep, that there would be room for nothing else in a typical postgraduate hydrology course. The cuckoo of the environment would push the legitimate hydrological fledgling out of the nest. Instead, hydrologists must become competent to assess environmental impact through the deepening and widening of their interests as hydrologists. It is not necessary, or possible, to become everything else, or even perhaps something else, in addition to being hydrologists. The better we are as hydrologists, the more accurately will we foresee the consequences of our works and proposals on the environment and if our general education is adequate we will accept the necessity to inform, or consult with, experts in other areas and to inform appropriate authorities. The solution lies not in the creation of a new science, profession or discipline and certainly not in joining a band wagon more notable for its enthusiasm than its scientific reserve, but in the deepening of the competence of professional hydrologists in their own domain, though greater acquaintance with the natural sciences and the scientific method.

References

Nash, J.E., 1973. Real Hydrology. Rep. No. 20, Institute of Hydrology, Wallingford, Oxon, pp. 17–25.

Nash, J.E., P.S. Eagleson, J.R. Philip and W.H. van der Molen, 1990. The education of hydrologists (Report of the IASH-UNESCO Panel). Hydrol. Sci. J., 35(b): 12.

Chapter 13

Hydrology and the real world

J.R. PHILIP

Abstract

This salute to Professor James Dooge celebrates his personal qualities, and goes on to discuss various problems and dichotomies in the fabric of present-day hydrology. Topics include: linearity and non-linearity; pseudophysical and regression models; the scientific-hydrology/professional-practice dichotomy; the discipline of the real-world phenomenon.

1. Introduction

Professor J.C.I. Dooge has chronicled the history of hydrology, pondered its philosophy, and wrestled with its theoretical basis. The history of the evolution of hydrology over the millennia from animist superstition to the beginnings of a soundly based geophysical science makes fascinating reading (Biswas, 1970; Philip, 1978a; Dooge, 1984); and Jim Dooge has contributed actively to present-day knowledge of this story. As early as 1959 he brought to attention the pioneering hydrologic efforts of Pierre Perrault (Dooge, 1959). He has treated a wide range of matters hydrologic. He has discussed the comparative planetology of water in the solar system, the consequences of water's molecular structure, and the disposition of water on Earth; and he has envisaged different views and insights about water which might be held by biologists, physicists, "social scientists", and indeed hydrologists (Dooge, 1983, 1984). The interested may find pertinent supplements on the physical chemistry, cosmology, geography, and biology of water in Fletcher (1978), Ringwood (1978), Philip (1978a), Robinson (1978), Macfarlane (1978), and Cowan (1978).

2. Hydrology and non-linearity

But I must address the topic of this Festschrift, theoretical hydrology. "Linear Theory of Hydrologic Systems" (Dooge, 1973) is perhaps Jim Dooge's *magnum opus*. This was indeed an ambitious essay in theoretical hydrology, in which were poured out the multitude of things hydrologic that Jim Dooge knew about, or worried about (in all of which he took obvious pleasure).

For me the epithet "linear" in the title was something of a turn-off. I had spent two decades struggling with the non-linearities of soil–water flow in unsaturated soils which are at the heart of terrestrial hydrology (Philip, 1954, 1969); and I was well aware also of the inherent non-linearity of the processes of overland flow (Philip, 1956) and of channel and river flow. But a careful reading of "Linear Theory" reveals that Dooge was perfectly aware of these non-linearities. In this he differs conspicuously from some contemporary linear enthusiasts, who fly in the face of physics and blandly proclaim linearity in data that shouts a contrary message. Jim Dooge may well recoil sadly from the perversion of his doctrines by latter-day followers.

I go on here to address (what in my view are) unresolved dichotomies and worrying deficiencies in present-day hydrology. I necessarily organize this in a linear sequence under separate heads, but none of these items should be taken in isolation. Each has close logical and psychological interconnections with most (and sometimes all) of the others.

First I mention in passing one dichotomy which (in my view) hydrologists seem reluctant to face: that between pseudophysical models, on the one hand, and regression-type models on the other. A discussion of this elsewhere at greater length (Philip, 1975) seems to be still pertinent. Pseudophysical models have the treacherous pitfall that supposedly physical parameters may be strongly correlated (either positively or negatively) so that parameter estimation is fraught with error. In these circumstances a regression model, with no pretensions that the parameters have physical meaning, but in which they are at least "orthogonal", may be more efficient. This alternative, of course, usually depends on a statistical assumption of stationarity. But non-stationarity is the absolute essence of the studies of anthropogenic change and of natural (e.g. climate) change which increasingly dominate the hydrologic agenda.

3. Natural science and trans-science

The term "natural science" embraces both the physical and the biological sciences and is in contradistinction to the "social sciences". Hydrology is the study of the totality of physical processes involved in the hydrologic cycle, on all scales, and in the interactions between them. Like the other branches of geophysics it is (or at least it should be) a natural science, subject to the criteria of natural science and advanced by its methods.

These are well described, for example, in the writings of Polanyi (1951), Bronowski (1951, 1961), Popper (1959), and Merton (1968). Natural science is directed towards the extension of certified knowledge, with knowledge understood as empirically confirmed and logically consistent predictions. There stems from it an ethos based on certain essential norms, one of which is organized scepticism: the scientist must ultimately accept nothing on trust, neither the work of others nor his own speculations. The sole test of any hypothesis (theory, model)is how well it is supported by accurate observations or measurements of the actual phenomenon in the real world. Work purporting to be scientific, but which infringes this norm, lacks effectiveness and integrity (Philip, 1978b, 1991).

An hydrology fully satisfying the norms of natural science is, of course, an ideal which, certainly for now, lies way outside our reach; and there is reason to believe that it will remain always partly unattainable.

The gaps in the natural-science fabric of hydrology must not, however, lead us into paralysis about the hydrologic problems of the practical world about us. As Ortega y Gasset (1944) said, "Life cannot wait until the sciences have explained the universe scientifically. We cannot put off living until we are ready." Pressing problems demand attention whether or not the relevant natural science is available.

It is important for hydrologists to recognize when they are operating on a sound scientific base, and when they are not. They should not delude themselves (and others) that their predictions have a solid natural-science imprimatur when they do not.

Natural science may fall short of providing answers to hydrologic problems in two distinct ways. Firstly, and more obviously, there are areas of hydrology which are amenable in principle to the methods of natural science, but for which the appropriate scientific attack has not yet been developed. That such areas exist in the juvenescent science of hydrology should evoke no surprise. The ongoing challenge to scientific hydrologists is to reduce these areas.

Beyond this, we must recognize a graver, and more insidious, limitation: there is a second, irreducible, class of hydrologic questions to which natural science can never provide certified answers. Such questions are "trans-scientific" in the sense of Weinberg (1972). Alvin Weinberg discussed the category of questions which "can be stated in the language of science [but which] are unanswerable by science." Many practical problems of hydrology are trans-scientific. Indeed we are deep into trans-science whenever we face a question which, for an accurate answer, demands a greater expenditure of resources (and/or time) for experimentation and/or data collection and/or subsequent computation, than can ever be justified or afforded (Philip, 1975).

Hydrologists have the responsibility to bring their expertise to bear on such trans-scientific questions, but when they do they stray into uncharted territories outside the well-ordered realm of certified knowledge. Our trans-scientific judgements should, of course, be informed and sustained by a scientific hydrology: a sceptical science with a coherent intellectual content based firmly on the real phenomena. But, as Dr Weinberg has said, "The most science can do is to inject some intellectual discipline into the republic of trans-science."

4. Scientific hydrology and professional practice

A major part of the educational effort in hydrology occurs in engineering schools and tends to be restricted to certain sub-fields (Nash et al., 1990). Various authors have attributed to this the fragmented nature of much hydrologic expertise and the slowness of progress towards an intellectually coherent scientific hydrology. For example, Dr V. Klemes has written that engineering students are conditioned to be piecemeal tacklers of individual problems in isolation, rather than natural scientists. The latter, on the other hand, he sees as curiosity-driven and motivated to contribute to a consistent total scientific hydrology.

The IAHS/UNESCO panel on hydrologic education (Nash et al., 1990), pondering the problem at some length, remarked *inter alia* that engineers coming to hydrology "may have tended to neglect the possibility of applying established scientific principles, and the necessity for careful observations. They may have tended to rely excessively on the mathematical analysis of 'data', i.e. observations, usually made by other persons, often for other purposes ... This approach, though providing many interesting challenges, which often left room for an almost Renaissance exercise of ingenuity in problem solving, was not, however, a sound basis on which to build an applied science." (That I prefer the adjective "medieval" over "Renaissance" here will be clear later.)

In my view the scientific-hydrology/engineering schism is part of a larger dichotomy between natural science and professional practice. The following paragraphs are a paraphrase from Philip (1991). Ortega y Gasset cannot be gainsaid; the world cannot grind to a halt until all the scientific answers are in (and, indeed, even if they were, vast and embarrassing trans-scientific gaps would remain). The world's professionals (engineers, agronomists, foresters, planners) cannot stand on scientific niceties: they must be about their business. Engineering schools epitomize education for a profession. The message of my engineering course in the 1940's was: "All things are understood, and all a young engineer needs to know is what handbook to use." (Today's update: "... what software to use.")

This will seem eminent good sense to many professionals. The professional must get on with the job, with minimum fuss and delay, and he is best able to do this without hesitation when he avoids personal responsibility (and, indeed, legal liability). Where there are laws, regulations, standard specifications, handbooks, rule-of-thumb equations, and extant software and models, there is a clear road ahead for him. Should the exercise turn out badly, no fault lies with him. He has merely obeyed the instructions of his bosses and the injunctions of his profession.

I need hardly remark that this mind-set is diametrically opposed to that of natural science (at least as an ideal) and is a most negative preparation for a research career. Many involved in hydrology and kindred fields drift uneasily between the two camps of natural science and professional practice, and it is not surprising that some become uncertain in their values and motivations.

I do not know how we can resolve this conflict. But it would be a beginning if the dichotomy were frankly recognized and comprehended, and if each camp were to understand the other's values and to accord to the other its essential rôle.

5. Hydrology and the real world

We have hinted at the tendency for certain types of hydrologist to stay remote from real-world phenomena. In the ancient one-liner, "Don't confuse me with the facts." The advent of high-speed computing has exacerbated the malady. Here again I paraphrase Philip (1991).

Of the making of models there is no end. The component parts may be well-founded in natural science, may be a crude simplification, or may be no more than a black box which has taken the modeller's fancy. Equally, the prescriptions for fitting the components together may or may not be well-based. Beyond the question of the validity of the model's machinery, there are also difficult questions of the quality both of the parametrization of the components of the model and of the data inputs.

A disturbing development is that computer modelling has largely supplanted laboratory experimentation and field observation as the research activity of both undergraduates and graduate students. Most raw PhD's seeking a job at our Centre proudly present themselves as computer jockeys, incurious about real-world phenomena and innocent of laboratory and field skills, yet blissfully unaware of their inadequacy for serious research.

The IASH/UNESCO panel (Nash et al., 1990) reported thus:

> "One urgent educational problem which has reached crisis proportions in many universities, is the lack of field and laboratory experience. This is a problem at all levels and in many disciplines and has existed for long enough to become self perpetuating through the next generation of faculty. The consequences in hydrology are both profound and disturbing especially with the current emphasis on conceptual modelling. Although such models constitute useful tools in the investigation of the physical world, exclusive or undue reliance on them may tend to separate students from the realities they are supposed to study. In the absence of appropriate testing, models take on an aura of reality in the minds of their users and become a source of unsound science and practice."

It is sad to see the young going down this track. They cannot be blamed. For them to develop otherwise demands a mental and moral integrity we cannot expect in young persons given open slather to play around with glamorous space-age toys.

6. Hydrology and faith

Thomas Aquinas (1989, p. 334), with his sure grasp of words, takes from the letter to the Hebrews his definition of faith: *Faith is the substance of what we hope for, the evidence of what we cannot see.* Natural science began from a minimal, but transcendental, crumb of faith: the hope that there was order in the world around us that we could find, understand, and use. The visible evidence of the success of science has, of course, long driven out any place for faith.

But all too much of hydrology has about it the air of medieval scholasticism. It exhibits an excess of faith, of reliance on "the evidence of what we cannot see", when the real evidence is observable and measurable out there in the real solid world. There is a glut of hope and a shortage of scepticism.

I expect Jim Dooge would agree.

Acknowledgments

This article was prepared during a visit to the Department of Engineering Hydrology, University College, Galway, and to the Centre for Industrial and Applied Mathematics, Oxford University. I am grateful to both institutions for financial help.

References

Aquinas, T., 1989. Summa Theologiae, a concise translation. T. McDermott (Editor). Eyre and Spottiswoode, London.

Biswas, A.K., 1970. History of Hydrology. North-Holland, Amsterdam.

Bronowski, J., 1951. The Common Sense of Science. Heinemann, London.

Bronowski, J., 1961. Science and Human Values. Hutchinson, London.

Cowan, I., 1978. Water use in higher plants. In: A.K. McIntyre (Editor), Water: Planets, Plants and People. Australian Academy of Science, Canberra, pp. 71–106.

Dooge, J.C.I., 1959. Un bilan hydrologique au XVIIe siècle. La Houille Blanche, 6: 799–803.

Dooge, J.C.I., 1973. Linear Theory of Hydrologic Systems. USDA ARS Tech. Bull. 1968, U.S. Govt. Printing Office, Washington.

Dooge, J.C.I., 1983. On the study of water. Hydrol. Sci. J., 28: 23–48.

Dooge, J.C.I., 1984. The waters of the earth. Hydrol. Sci. J., 29: 149–176.

Fletcher, N.H., 1978. Water — a unique fluid. In: A.K. McIntyre (Editor), Water: Planets, Plants and People. Australian Academy of Science, Canberra, pp. 4–34.

Macfarlane, W.V., 1978. Ecophysiology of water in desert organisms. In: A.K. McIntyre (Editor), Water: Planets, Plants and People. Australian Academy of Science, Canberra, pp. 108–143.

Merton, R.K., 1968. Social Theory and Social Structure. Free Press, New York, N.Y.

Nash, J.E., Eagleson, P.S., Philip, J.R. and van der Molen, W.H., 1990. The education of hydrologists. Hydrol. Sci., 35: 597–607.

Ortega y Gasset, J., 1944. The Mission of the University. Princeton University Press, Princeton, N.J.

Philip, J.R., 1954. Some recent advances in hydrologic physics. J. Inst. Eng. Aust., 26: 255–259.

Philip, J.R., 1956. An application of the diffusion equation to viscous motion with a free surface. Aust. J. Phys., 9: 570–573.

Philip, J.R., 1969. The theory of infiltration. Adv. Hydrosci., 5: 215–296.

Philip, J.R., 1975. Some remarks on science and catchment prediction. In: T.G. Chapman and F.X. Dunin (Editors), Prediction in Catchment Hydrology. Australian Academy of Science, Canberra, pp. 23–30.

Philip, J.R., 1978a. Water on the Earth. In: A.K. McIntyre (Editor), Water: Planets, Plants and People. Australian Academy of Science, Canberra, pp. 35–59.

Philip, J.R., 1978b. Towards diversity and adaptability: An Australian view of governmentally supported science. Minerva, 16: 397–415.

Philip, J.R., 1991. Soils, natural science, and models. Soil Sci., 151: 91–98.

Polanyi, M., 1951. The Logic of Liberty. Routledge, London.
Popper, K., 1959. The Logic of Scientific Discovery. Hutchinson, London.
Ringwood, A.E., 1978. Water in the solar system. In: A.K. McIntyre (Editor), Water: Planets, Plants and People. Australian Academy of Science, Canberra, pp. 18–34.
Robinson, J.R., 1978. Water in the animal body. In: A.K. McIntyre (Editor), Water: Planets, Plants and People. Australian Academy of Science, Canberra, pp. 60–70.
Weinberg, A.M., 1972. Science and trans-science. Minerva, 10: 201–222.

Chapter 14

Water and soil: circulation and risks of pollution

GEORGE VACHAUD and MICHEL VAUCLIN

Abstract

The growing problem of ground water pollution from diffuse sources, associated with land use and agricultural practice, requires the cooperation of specialists from a wide range of disciplines. The key to removing barriers between these disciplines is to focus on understanding the flow of water and its interactions in the soil, as a hydrological system at different spatial scales. We review the progress that has already been made and the path that lies ahead.

1. Introduction

One of the facts that mark this past decade has been the recognition of the fragility of our natural environment and the limitation of our available resources. Among these, water holds an essential place: it is indispensable, as much for human and animal life as for the development of plants. We have been blessed with rain, for a long time free of charge and worry, until recent climatic variations demonstrated the risk of drought and the cost of its availability. Of all the renewable resources, it is probably the one which is most sensitive to all sorts of contamination.

The soil and the sub-soil form a very important hydraulic reservoir wherein water is stored, after travelling across the surface of the soil. Starting with a good understanding of physical, chemical and biological processes, which intervene between water and soil during these transfers, one may be able to predict the changing quality of underground water resources. Water present in the soil is also one of the essential components of agriculture. Crops need water, mineral and organic compounds close to their roots, for their development. Certain elements are also potential pollutants, particularly for groundwater, e.g. nitrates and pesticides. This necessarily leads to a conflict of use between agriculture and the supply of water for human needs.

These two aspects "quality" and "quantity" are inexorably linked whenever one discusses the theme of water and soil. In this paper, we shall attempt to present and illustrate certain problems which arise in the evolution of available groundwater resources as a function of agricultural practices.

2. Presentation of the problem

Even though it is beneath our feet and hidden from view, which is not the least of our problems at the level of quantification of resources and transfers, groundwater represents one of the essential components of the hydrological cycle.

This underground reservoir may be divided into two zones:

– *The zone of aeration or the unsaturated zone* which extends from the surface of the soil to the level of the groundwater table. This is the critical zone through which water moves before entering the underground aquifer. The root system of plants also develops within this zone. We also find there essential microbiological processes and the assimilation of the major elements of transformation.

– *The saturated zone of aquifers* which serves as an important reserve of water. For example, in France approximately 30% of the water used for industrial, agricultural and industrial consumption is provided by underground aquifers. This proportion rises to 80% in Germany and in the Netherlands.

This underground reservoir is supplied by rain and by the surface reservoirs: rivers, lakes, ... In an average year, with a rainfall of 800 mm, the quantity of water which infiltrates into the soil is approximately 600 mm, or alternatively, 6 million m^3/km^2. Approximately 400 mm are recovered by the vegetation as evapotransformation. The net supply to aquifers represents approximately 200 mm. It is essential to maintain an equilibrium between supply to aquifers and extractions. These numbers may be considered simply as orders of magnitude, since they vary as a function of climate, soil and vegetation.

During the circulation of the water in the interstices of the soil with its characteristic porosity and through a zone within which the flow velocities may be very weak, the water comes in contact with all sorts of substances fixed in the solid matrix or dispersed in the soil after absorption on the soil surface. If we consider trace elements integrated in the geochemical cycles, for example, the dissolution of calcium and sulphate by soil water, the exchanges following the entrainment of dissolved particles have a limited effect on the quality of groundwaters. The consumers of mineral water are particularly appreciative of them also. It is a completely different story when we are concerned with substances resulting from agricultural practices, such as, application of fertilizers, animal slurry, herbicides and pesticides, industrial discharges, accidents, leakage of hydrocarbons and radioactive substances... Minute traces for example 0.5 microgrammes per litre for pesticides, may represent levels of pollution which are judged to be dangerous. The understanding of the transfer mechanisms governing water moving through soil, and of the interactions between the water–soil system and dissolved elements, is therefore essential in estimating the mass concentration of pollutants arriving at the water table. Water constitutes the essential vector for the migration of pollutants at depth.

3. Circulation of water in the soil

The study of the different transfer mechanisms is complicated by the fact that it is necessary to consider interacting phenomena in a polyphase (water–air–solid) deformable medium which is the site for changes of phase and physico-chemical exchanges, and is subject to random boundary conditions. When one begins to study transfer mechanisms in such a medium, three scales of observation may be considered.

3.1. The microscopic scale of the pore

At this scale, the magnitude of local quantities exhibit large fluctuations. Each phase may be considered as a continuous medium. We may regard the soil as a set of continuous, more or less interpenetrating media. This is the scale for understanding the fundamental mechanisms which come into play.

3.2. The macroscopic scale: the laboratory column

The complex geometry of the pore space makes it very difficult to pose a precise definition of the topology of the different phases and to measure the associated variables and parameters. This ensures that the microscopic description cannot be used in practice without the change of scale that leads to the introduction of the concept of Representative Elementary Volume — REV. This REV permits the establishment of an equivalence between the real dispersed medium and a fictitious continuous medium. At this scale the phenomena are described by equations containing variables and parameters which are representative of average quantities defined on a continuous porous medium.

Two essential quantities intervene in the flow mechanism of water in the soil:

– *The volumetric moisture content*, defined as the volume of water present in a volume of soil. It varies between two limits: zero when the soil is dry, and the full porosity, when all the interstices of the soil (from 30 to 50% of the total volume) are filled with water. Its measurement permits, among other things, the definition of the volume of water present in the soil and the variations induced by extraction through roots, by evaporation through the surface of the soil, by percolation towards the water table or by infiltration of water or irrigation.

– *The pressure of water*, which characterises the importance of surface forces between the water and the porous matrix. It represents the energy required to transform the soil from one soil–water state to another. This quantity is related in a complex fashion to the water content of the soil through the retention, or suction curve, the form of which depends very strongly on the nature, structure and texture of the soil.

Finally, the kinetics of mass transfer in soil is essentially controlled by the relationship between the hydraulic conductivity of the soil and its content of water. This relationship has the peculiarity of being very strongly non-linear, the conductivity decreasing more rapidly than the content. In addition this relationship depends on the distribution of solid soil particles and of the topology of the pore space. Consequently, the kinetics of water transport in soils are extremely complicated. Similarly, if one considers the infiltration of a volume of water following its placement as a layer approximately 1 cm thick on the surface of a column of fine sand or of clay loam of 1000 cm^2 in section, we find:

- in the first case, it takes 40 minutes for a ten litre volume of water to infiltrate, wetting the column to a depth of 60 cm;
- in the second case, it takes 5 days to infiltrate the same volume of water and the depth wet is 40 cm.

From the formal point of view the equation which describes the flow is obtained by combining the equation of conservation of mass:

$$\frac{\partial \theta}{\partial t} = -\operatorname{div} q \tag{1}$$

with the equation of Darcy generalised to unsaturated media, i.e. taking account of the non-linearity between hydraulic conductivity and the moisture content, namely:

$$q = -K(\theta) \operatorname{grad} (h - z) \tag{2}$$

where z is the vertical axis orientated positively downwards, q is the volumetric flux density, θ is the volumetric moisture content, h is the pressure of soil water with respect to atmospheric pressure expressed in terms of the height of a column of liquid, and K is the hydraulic conductivity.

In the case of one-dimensional vertical flow and using the hypothesis of a single-valued relationship between h and θ (the case of pure infiltration or pure drainage), the equation of flow may be written in the form of equation (3), the equation of Richards:

$$C(h) \cdot \frac{\partial h}{\partial t} = \left[K(h) \cdot \left(\frac{\partial h}{\partial z} - 1\right)\right] \tag{3}$$

which describes the transfer of water in a vertical column of soil. $C(h)$ is the capillary capacity $C = \mathrm{d}\theta/\mathrm{d}h$, which takes account of the capacity of an unsaturated soil to store or to release water following a change in pressure. Writing this equation in terms of the pressure h as the dependent variable, presents a unique advantage in hydrology. This form permits in effect the coupling between flow in the unsaturated medium (h less than 0) and the flow in a saturated medium, below the groundwater table (h greater than 0). Equation (3) changes to the Laplace equation when h is greater than or equal to 0 and $\theta = \theta_{sat}$ (Vauclin et al., 1976). Due to the strongly non-linear character of $C(h)$ and $K(h)$ in equation (3), only the methods based on numerical analysis are applicable in the general case. A number of analytical solutions have however been developed for very particular cases:

- an initial, uniform profile, with infiltration under constant pressure and close to zero (Philip, 1969; Parlange et al., 1985);
- an initial, uniform profile, with infiltration under constant flux (Parlange, 1971);

- an initial, variable profile, with infiltration under constant flux (Boulier et al., 1984).

Many discrete numerical schemes have been proposed. Vauclin et al. (1979) carried out a systematic study of the choice of different schemes and weighting modes. This led to the proposal of an implicit scheme with explicit linearisation and the estimation of inter-nodal hydraulic conductivities by taking a geometric average of the values at the nodes of the calculation.

3.3. Megascopic scale: the landscape

An additional parasitic effect makes the analysis of problems even more complicated at the megascopic scale of the landscape: the spatial variability of the environment. Cultivated soils, in effect, consist of matter of sedimentary and aeolian origin, characterised by the existence of zones of accumulation in the micro-relief of the landscape. This leads to the presence of lenses and layers, usually of small dimension, but which may locally have considerable effect on the kinetics of flow and the transport of chemical substances. These are not easily detectable by pedological prospecting. For example, an intensive experiment was carried out in Senegal in order to characterise the water demand of peanut plants at the scale of 1 hectare, in an environment reputedly very homogenous. Considerable differences were measured a few metres apart in the storage of water in the soil and in the depth of infiltration by rain. This last quantity varied between 1 and 3 metres.

One may note that an essential characteristic of the variability of mass transfer parameters in the soil, is the existence of very strong spatial autocorrelations, depending essentially on the mode of formation of the soils. In order to avoid statistical bias (Vieira et al., 1983), it is necessary to take into account the spatial dependence of the data, whether for the interpolation of measurements (Delhomme, 1976; Webster and Burgess, 1980; Munoz-Pardo, 1987), or for the development of predictive models.

The concept of geometric similitude of porous media has been used in the purely stochastic modelling of hydraulic transfer at the megascopic scale (Miller and Miller, 1956). This depends on expressing the spatial variability of hydrodynamic properties using a scale factor which is treated as a random variable (Nielsen et al., 1973; Warrick et al., 1977; Simmons et al., 1979). Various stochastic solutions (analytic or numeric) based on this concept have been proposed (Vauclin and Vachaud, 1984) and applied with success in the case of infiltration and drainage of heterogenous soils.

It is therefore evident that the estimation of hydraulic transfers and their modelling, demands, not simply the determination of the hydrodynamic parameters at a point, but also (a) the characterisation of their variability at the scale of the study (field, catchment), and (b) the development of models of the stochastic type that take into account the uncertainties depending on these spatial fluctuations, in the presence, generally, of an insufficient number of observations.

4. Transport of dissolved substances

In numerous cases in agriculture, the flow of water in the soil is accompanied by the entrainment of dissolved substances in association with root-extraction and an imbalance between the supply to the plant and its needs. Because of this fact, agriculture may contribute to the pollution of the underground environment, and moreover, it may take place over large areas; hence the term diffuse pollution, in contrast to industrial pollution. For example, in the plain of Alsace, the Agence de Bassin Rhin-Meuse, estimates that between 50 and 60% of the degradation in the quality of groundwater is due to agricultural activities.

The presence of chemical elements in the soil brings into play a series of elementary processes that are relatively simple to analyse when considered in isolation. However, these become globally very complicated because of numerous interactions between individual processes, the complexity of the porous matrix, the presence of air in variable concentration in the medium, and finally the transitory character of flows that often lead to thermodynamic phenomena far from equilibrium. These include adsorption and desorption on solid surfaces, microbiological and physico-chemical transformations and reactions, and ionic exchanges.

As an example, let us consider the transformations of nitrogen and pesticides, which constitute two classes of problem particularly important at the present time (see Figure 1). Following the application of nitrogen fertilizer, part of the nitrogen is taken up by the plants, and the rest is transformed into organic nitrogen by the micro-organisms present in the soil. This organic nitrogen is then transformed into

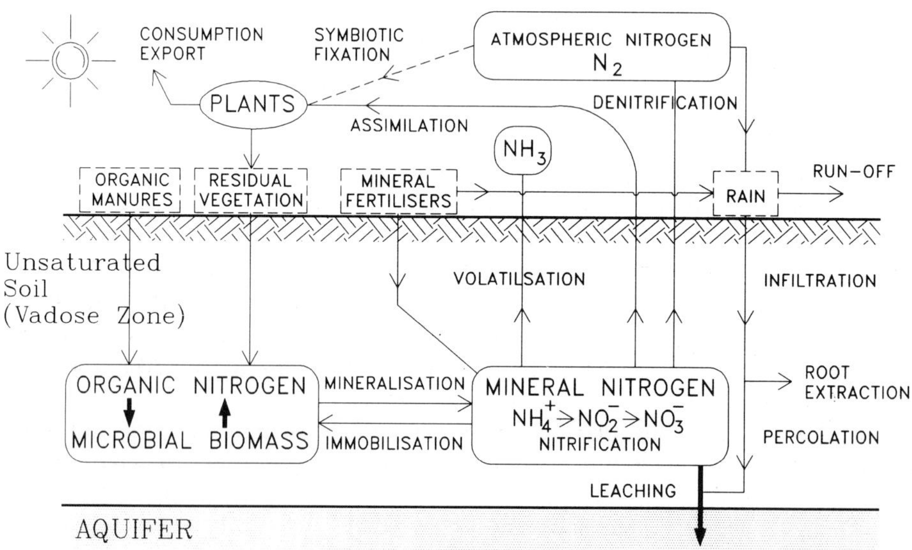

Fig. 1. The nitrogen cycle.

ammonium by ammoniacal bacteria, and subsequently adsorbed by the clay-humic complex of the soil. The ammonium may be transformed into nitrite and then into gaseous nitrogen that returns to the atmosphere. The nitrite ions, because of their solubility, are also susceptible to very rapid leaching by percolating water and entrainment by convection towards the water table.

In agriculture, an over-estimation of the dose of nitrogen fertilizer to be applied, is therefore a risk factor. The error may occur either in the method of application or in the frequency of application. An application rate greater than a few 100 kg/ha per annum is of questionable value to agriculture, both from the point of view of its effect on production and of the risk of pollution. Rates of application frequently exceed this amount by 30 to 50% in order to guarantee a return.

It is clear that the major risks occur in Autumn and in Winter because of the coincidence of heavy rainfall and the absence of vegetation cover. Crops such as maize, tobacco and root crops, should be subject to particular surveillance. These are generally planted in free-draining soils that require both the supply of large volumes of water by irrigation and strong fertilisation, and leave the soil bare during long periods.

Another set of pollutants which are much more dangerous from the point of view of human health, even in very small concentrations, are the organo-chlorides: pesticides or herbicides. The study of the transfer of these substances (see Figure 2) is much more complicated than in the case of nitrogen. These compounds and their reactants, their solubility in water, their absorption by clay particles and organic matter of the soil, their degree of mobility in the liquid phase, and their degradability in the environment are controlled by very different time constants and parameters.

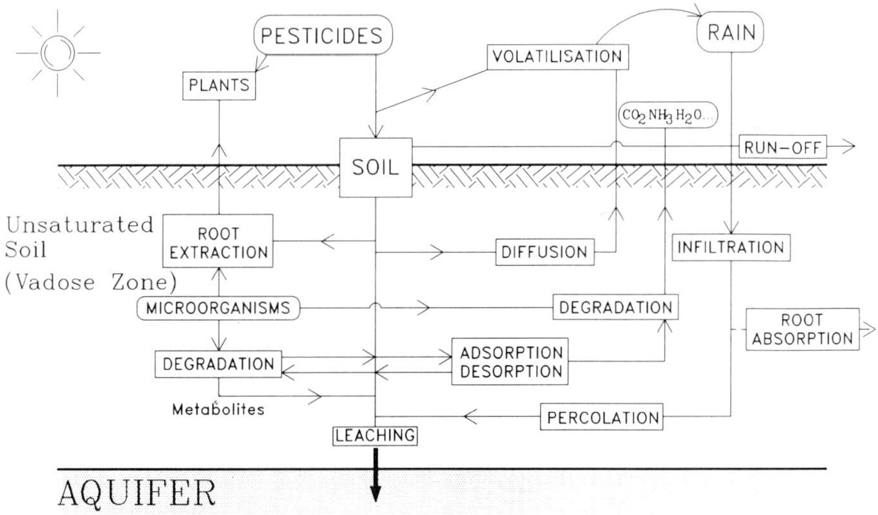

Fig. 2. The cycle of xenobiotic substances.

Generally, in every case, the elementary processes are of the same type as those for nitrogen: degradation by micro-organisms in the soil, leading to less toxic mineral metabolites, volatilisation in the gaseous phase, adsorption–desorption by the solid phase of the soil.

In both cases, the essential point of difficulty at the level of understanding the individual process, is the role of micro-organisms. One gramme of soil may contain between ten million and one milliard organisms. Each one has a specific function and environmental conditions or requirements: humidity, temperature etc. The step from the microscopic study of these elementary processes to the prediction of their consequences at large scale (many square kilometres) will remain, for a long time, a challenge for soil science.

At the level of the formal description of the transfer of dissolved substances of the soil, it is clear that in the majority of cases, convective entrainment by water represents the most important phenomenon. The transfer equation which applies, as in the case of hydraulic transport, is derived from two principles:

- *the equation of conservation of solute mass*:

$$\frac{\partial}{\partial t}(\rho S + \theta C) = -\frac{\partial J_s}{\partial z} + \sum \Phi_j \tag{4}$$

where J_s is the mass flux density of solute, S and C are the concentrations of solute associated with the solid and liquid phases, respectively, Φ_j is the rate of creation and destruction of the solute j in the medium, and ρ is the volumetric mass of soil;

- *the equation of transfer*:

$$J_s = -\theta \cdot D_s \frac{\partial C}{\partial z} + qC \tag{5}$$

where D_s is the coefficient of hydrodynamic dispersion.

The insertion of equation (5) in equation (4) leads to the equation for transfer in the general form:

$$\underset{(a)}{\frac{\partial \rho S}{\partial t}} + \underset{(b)}{\frac{\partial \theta C}{\partial t}} = \underset{(c)}{\frac{\partial}{\partial z}\left(\theta \cdot D_s \frac{\partial C}{\partial z}\right)} - \underset{(d)}{\frac{\partial}{\partial z}(qC)} + \underset{(e)}{\sum \Phi_j} \tag{6}$$

It is clear that this equation contains several coupled and generally complex processes. The term (e) represents essentially the processes of chemical reaction (precipitation dissolution, oxidation and reduction), or biochemical reactions (katabolic reactions and microbial transformations) as well as the absorption of solute by the roots. This is an ambiguous term representing processes that we do not know how to model properly and demanding generally the use of empirical expressions. Similarly, the term (a), which characterises the physico-chemical reactions at the solid–liquid interface by means of a rate of exchange, is open to considerable debate, particularly concerning the question whether one may apply the hypothesis of local equilibrium or not, in treating the processes of absorption and desorption.

Finally, the coefficient of dispersion D_s in term (c), is nothing more than a method of representing the complex distribution of dissolved particles in the topology of the porous medium. As with many other parameters present in this equation, it is a term which does not have a real physical meaning and cannot be measured directly.

In many cases this approach is made even more complicated by the possible existence of zones of dead water in the porous medium, leading to a partition between mobile and immobile liquid with exchanges between the two phases (Coates and Smith, 1964; Gaudet et al., 1977).

In addition to the essential problem of the parameters used in the mass transfer model, the method of numerical discretisation chosen to solve equation (6) coupled with the equation of hydraulic transport, may lead to an important bias in the form of numerical diffusion. A systematic review of problems raised by the nature of the models proposed in the literature, has been presented (Vachaud et al., 1988) on the occasion of the colloquium "Non Point Source Pollution in Agriculture" organised by the U.S. Department of Agriculture at Logan, Utah.

5. Conclusion

In spite of the very notable progress achieved in the past twenty years, it appears to us, that the scientific community concerned with these problems, has a tendency to limit the domain of investigation and vision to isolated individual disciplines. It is imperative in the years to come, that the barriers should be removed between practitioners and theoreticians, between fluid-mechanists, physical-chemists, soil microbiologists, plant physiologists and agronomists, and between those charged with laboratory and field investigations.

This requires conciliation and reconciliation in order to achieve a single-minded approach to understanding the flow of water and its interactions in the soil as a hydrological system in a global context.

References

Boulier, J.F., Touma, J. and Vauclin, M., 1984. Flux-concentration relation-based solution of constant-flux infiltration equation: I. Infiltration into non uniform initial moisture profiles. Soil Sci. Soc. Amer. J., 48: 245–251.

Coats, K.H. and Smith, B.D., 1964. Dead and pore volume and dispersion in porous media. Soc. Petrol. Eng. J., 4: 73–84.

Delhomme, J.P., 1976. Application de la théorie des variables régionalisées dans les sciences de l'eau. Thèse Docteur-Ingénieur, Université Pierre et Marie Curie, Paris.

Gaudet, J.P., Jegat, H., Vachaud, G. and Wierenga, P.J., 1977. Solute transfer with exchange between mobile and stagnant water through unsaturated sand. Soil Sci. Soc. Am. J., 41: 665–671.

Miller, E.E. and Miller, R.D., 1956. Physical theory for capillary flow phenomena. J. Appl. Phys., 27: 324–332.

Munoz-Pardo, J., 1987. Approche géostatistique de la variabilité spatiale des milieux géophysiques. Thèse Docteur-Ingénieur, Grenoble, 251 pp.

Nielsen, D.R., Biggar, J.W. and Erh, K.T., 1973. Spatial variability of field measured soilwater properties. Hilgardia, 42: 215–260.

Parlange, J.Y., 1971. Theory of water movement in soils — one dimensional infiltration. Soil Sci., 111: 170–174.

Parlange, J.Y. and Braddock, R.D., 1980. An application of Brutsaert's and optimization techniques to the non linear diffusion equation. Soil Sci., 129: 129–145.

Parlange, J.Y., Haverkamp, R. and Touma, J., 1985. Infiltration under ponded conditions. Part I. Optimal analytical solutions and comparison with experimental observations. Soil. Sci., 139: 305–311.

Philip, J R., 1969. Theory of infiltration. Adv. Hydrosci., Academic Press, New York, N.Y., 5: 215–305,

Simmons, C.S., Nielsen D.R. and Biggar J.W., 1979. Scaling of field measured soil water properties. Hilgardia, 47, 77–173.

Vachaud, G., Vauclin, M. and Addiscott, T., 1988. Solute transport in the vadose zone: a review of models. Proc. Int. Symp. on Water Quality Modelling. LOGAN, Utah, Part 1, pp. 82–104

Vauclin, M. and Vachaud, G., 1984. Transferts hydriques dans les sols non saturés, non homogènes. Ann. Mines, pp. 63–74.

Vauclin, M., Khanji, J. and Vachaud, G., 1976. Étude expérimentale et numérique du drainage et de la recharge des nappes à surface libre avec prise en compte de la zône non saturée. J. Méc., 15: 1–42.

Vauclin, M., Haverkamp, R. and Vachaud, G., 1979. Résolution numérique d'une équation de diffusion non linéaire. Application à l'infiltration de l'eau dans les sols non saturés. Presses Universitaires de Grenoble, 183 pp.

Vieira, S.R., Vachaud, G., Nielsen, D.R. and Vauclin, M., 1983. The use of cokriging with limited field soil observations. Soil Sci. Soc. Am. J., 47: 175–184.

Warrick, A.M., Mullen, G.M. and Nielsen, D.R., 1977. Scaling field measured soil hydraulic properties using a similar media concept. Water Resour. Res., 13: 355–362.

Webster, R. and Burgess, T.M., 1980. Optimal interpolation and isarithmic mapping of soil properties — changing drift and universal kriging. J. Soil Sci., 31: 505–524.

Chapter 15

Global change, a catalyst for the development of hydrologic science*

PETER S. EAGLESON

Abstract

The scientific issues of global change have highlighted the importance and scale of the role of water in the earth system. A case is made for the need to recognize and pursue hydrologic science as a separate and distinct geoscience that cuts across the traditional atmospheric, oceanic, and solid earth sciences (Figure 1).

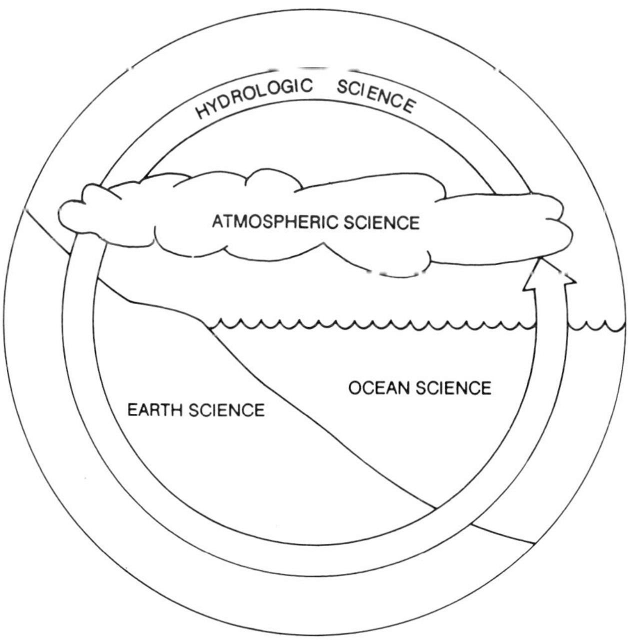

Fig. 1. Hydrologic science: a unifying geoscience (from NRC, 1990).

---

* This paper was presented as the Robert E. Hortin Lecture in the Global Change Symposium at the AMS Annual Meeting in Anaheim, California on 7 February, 1990.
Reprinted from *Bulletin of the American Meteorological Society*, Vol. 72, No. 1, January 1991, with permission of the American Meteorological Society.

1. Introduction

The scientific origins of hydrology lie in the engineering problems of flood control, water supply, and sanitation. Thus the education and motivation of its practitioners has been oriented to problem solving. Over time, with development in the size and complexity of human society, there has been an evolutionary expansion in the scale of these problems, but until relatively recently, scientific understanding has not been an organizing principle of the accompanying hydrologic research and education. While we have spent lavishly to cope with the scarcities and excesses of water and to ensure its potability, we have invested relatively little in the basic science underlying water's many roles in the planetary mechanisms.

The issues of global change have served to catalyze our interest in hydrologic science by highlighting the centrality of water processes to the physical climate system and to the biogeochemical cycles. With this has come the realization that the comprehensive hydrologic science base for understanding and coping with these issues is missing. Hydrologic science has a natural place as a geoscience cutting across the established atmospheric, ocean, and (solid) earth geosciences (see Figure 2), and its appropriate scientific research and educational infrastructure must be identified and established.

It is important to understand that what we propose here are not suggested as substitutes for or as pejorative reflections upon existing research or educational activities in the applied forecasting (i.e., engineering hydrology) and management (i.e., water resources) aspects of water. They are intended as needed complimentary, new initiatives.

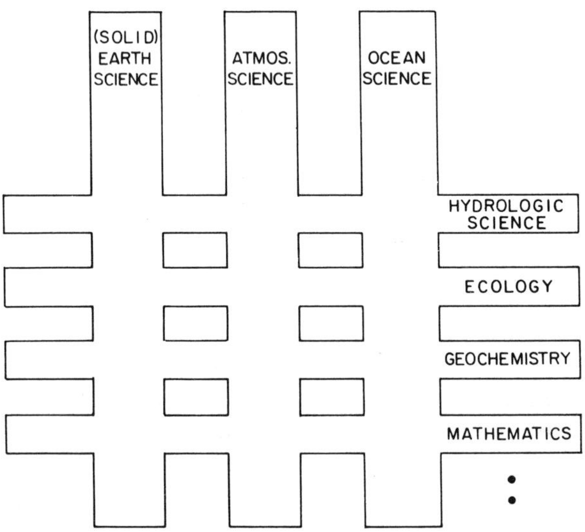

Fig. 2. The intellectual fabric of earth system science (from NRC, 1990).

2. Historical development of hydrologic science

Concern for water as both a necessity and a possible hazard has been with humans throughout their existence on earth, but not until the Hellenic civilization, about 600 B.C., did man attempt to understand nature just for the sake of understanding.

Precipitation was first measured in the fourth century B.C. by Kautilya of India and streamflow by Hero of Alexandria in the first century A.D. Little further advance in understanding occurred until the Renaissance when the work of Palissy (a French potter and naturalist), Perrault (a French lawyer), and Halley (English Astronomer Royal) collectively established the hydrologic cycle and marked the beginning of hydrologic science.

For the rest of the 17th and 18th centuries, Europe was transformed by the Industrial Revolution and its accompanying urbanization. Physicists and members of distinguished families lost interest in hydrology, and its development was left to civil engineers such as de Pitot, Chézy, and Venturi who were concerned with water supply and water power. The science developed spottily in response to the needs of engineering practice as water scientists and engineers focused their attention on drainage basins of characteristic dimension 10 to 100 km. Because the early foundations of hydrologic science were built upon experience with middle latitudes, some inadvertent and long-lived biases were established. The catchments being studied were assumed to be passive participants in the hydrologic cycle producing no feedback to the atmospheric forcing, and the hydrologic processes peculiar to the nondeveloping desert, tropical, and cold regions received little or no attention.

In the United States, hydrologic research remained the province of enterprising inventors, prospectors, and wealthy amateurs until late in the 19th century. At that time government became the prime mover in water research to support the practical needs of the proliferating water management projects on large rivers. New public agencies were formed such as the Weather Bureau and the Geological Survey. The early history of the United States Geological Survey (USGS) embodies the development of hydrologic science in the United States, particularly in sediment transport, groundwater, and water chemistry. (In surface water, however, private consulting engineers remained in the forefront of research well into the 20th century.) This period of federal and state agency dominance in the United States ended with the completion, in the 1950's, of the water projects delayed by World War II, and with the concurrent rise of the environmental movement.

The first formal recognition of the scientific status of hydrology was perhaps the formation of a Section of Scientific Hydrology within the International Union of Geodesy and Geophysics (IUGG) at its Rome assembly in 1922. At that time, the U.S. National Research Council's Committee for the IUGG was called the American Geophysical Union (AGU), and it was asked to add a section of Scientific Hydrology to its growing set of geophysical specialties. The fate of this proposal illustrates the status of hydrologic science within the U.S. scientific community during the first half of the 20th century. The leadership of AGU maintained for eight years that the level of active scientific interest (as opposed to consulting engineering interest) in the United States did not justify a separate section of scientific hydrology (Whitten, 1989).

In 1930, as AGU was transforming to an independent society, a Section of Hydrology was finally formed with O.E. Meinzer as chairman and R.E. Horton as vice-chairman. In 1931, Horton presented a comprehensive analysis of "The Field, Scope and Status of the Science of Hydrology" in which (1) the central focus was on the conservation of water mass at river basin scale, (2) concern was principally with the physics of the hydrologic cycle omitting any mention of chemistry, and (3) the effects of man were excluded by implication. These restrictions reflect the continued dominance of the field in the United States at that time by the engineering concerns of nation building.

In the meantime, beginning in 1926, the National Research Council had undertaken the preparation of a series of volumes on the physics of earth intended "to give the reader, presumably a scientist but not a specialist in the subject, an idea of its present status, together with a forward-looking summary of its outstanding problems". In 1936, upon the recommendation of the AGU, the NRC appointed a subcommittee on hydrology with O.E. Meinzer (geologist in charge, Division of Ground Water, USGS) as chairman to prepare a volume on hydrology as a conclusion to this series. Meinzer (1942) opened that volume with a definition that somewhat tentatively proclaimed hydrology to be an earth science but one with traditional methods and problems that set it apart. His definition went beyond Horton's, however, to incorporate water-related chemical and biological activity.

By the mid-twentieth century research on the scientific aspects of hydrology was well underway in university and government laboratories, focused on understanding laboratory-scale physical processes of the hydrologic cycle. Within the United States, concern was beginning to rise about the quality of our waters and about preservation of our natural environment. These enlarged interests found expression in an expanded definition of hydrology as a science put forward in 1962 by a committee chaired by Walter B. Langbein (Ad Hoc Panel on Hydrology, 1962). This definition included the chemical constituents of water and the relation of water to living things. The Langbein committee observed further that hydrology is an interdisciplinary science, involving an integration of other earth sciences to the extent that these help to explain the life history of water and its chemical, physical, and biological constituents. Missing here, however, is a sense of the *interactive* role of water in the earth system.

The U.N.-sponsored International Hydrological Decade from 1965 to 1974 raised consciousness about regional and global problems and about man's impact upon the hydrologic cycle. It became clear that man's changes to the landsurface were now affecting influential atmospheric processes and assessing these interactions was missing. It was the dramatic color photographs of earth in space, however, that crystallized active interest in the interconnectedness of nature and in the changes being wrought by man.

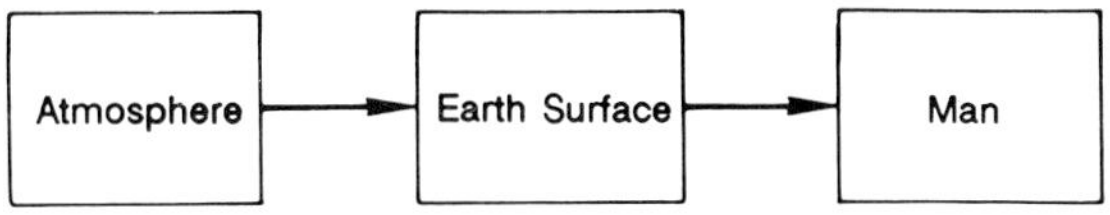

a. Classical Viewpoint

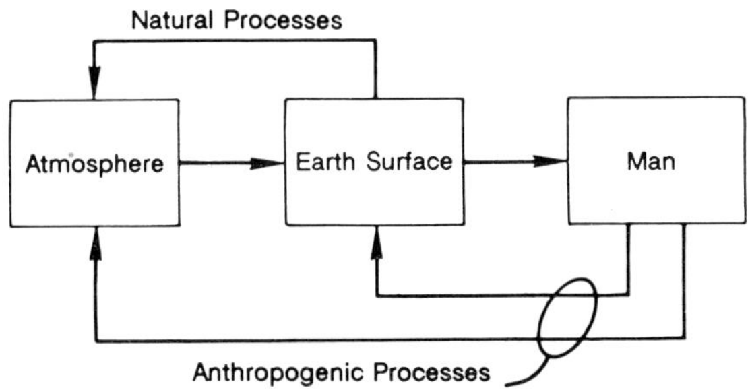

b. Modern Viewpoint

Fig. 3. The role of humans in the hydrologic cycle (from NRC, 1982).

Contemporary views of hydrology accept that human activity has become an integral and inseparable part of the hydrologic cycle (see Figure 3) and that the quality of water is no less a concern than the quantity. Furthermore, these views stress the dynamic interaction of water with the other components of the Earth system. It is these dynamic interactions of the hydrologic cycle with major physical, chemical, and biological processes across a wide range of space–time scales that suggest hydrologic science as a central science of global change.

3. The geophysical basis of hydrologic science

The global reservoirs and fluxes of water are shown schematically in Figure 4 in a representation of the hydrologic cycle as a fundamental geophysical process. Powered by the sun, the phase changes of water on earth involve storage and release of latent heat which helps drive the atmospheric circulation. Condensing in the atmosphere where it releases its latent heat, the liquid (or solid) water falls as precipitation, runs to the sea, and through evaporation regains its cargo of latent heat and returns through the atmosphere to wash and the land again. This climatic hydrologic cycle is the framework of hydrologic science and occurs over a wide spectrum of space and time scales. It is usually referred to simply as "the hydrologic cycle" and is illustrated quantitatively at global scale in Figure 5. It affects the global

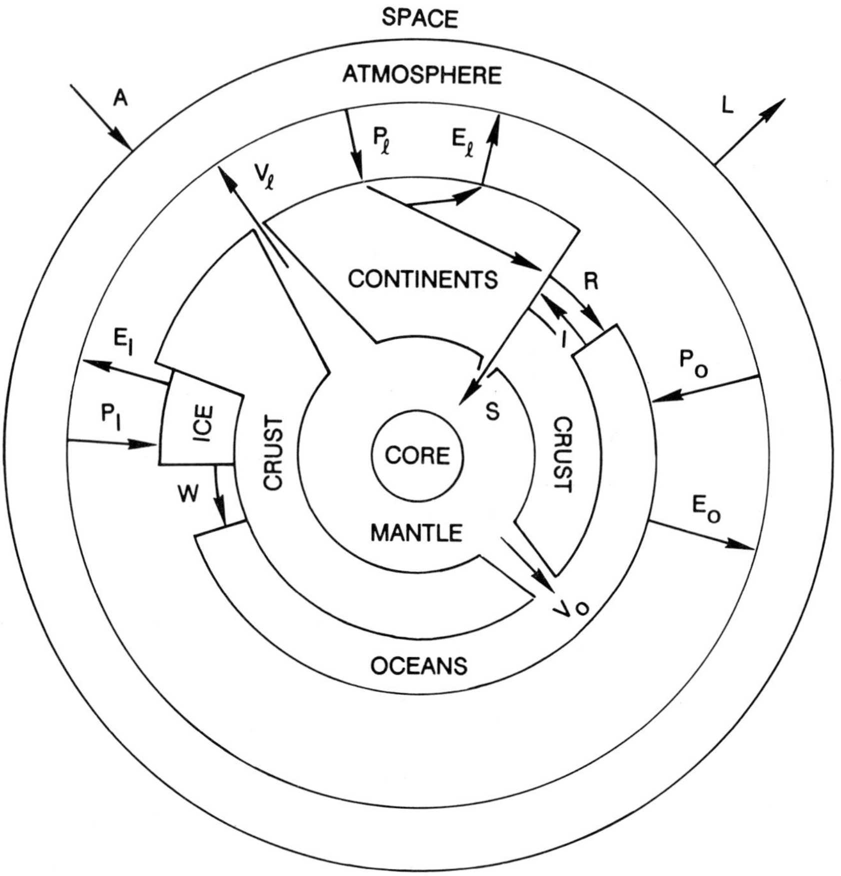

A	=	Additions of water from space
E_o	=	Evaporation from oceans
E_I	=	Evaporation (i.e., sublimation) from ice
E_ℓ	=	Evapotranspiration from land
I	=	Intrusion of seawater into continental aquifers
L	=	Loss of water to space
P_o	=	Precipitation on oceans
P_I	=	Precipitation on ice
P_ℓ	=	Precipitation on land
R	=	Runoff from continents
S	=	Subduction of water-containing crust
V_o	=	Volcanic venting to oceans
V_ℓ	=	Volcanic venting to atmosphere
W	=	Wastage of ice sheets to ocean

Fig. 4. The hydrologic cycle as a global geophysical process (from NRC, 1990).

circulation of both atmosphere and ocean, and hence is instrumental in shaping weather and climate.

There is another hydrologic cycle important to global change, but at geologic time scales, in which movements of the earth's crust and mantle (rather than atmosphere and oceans) are the agents for water transport. In this cycle, water becomes incorporated chemically into rock (for example, by oceanic sedimentation of animal shells), subducted to the mantle through plate movement and returned

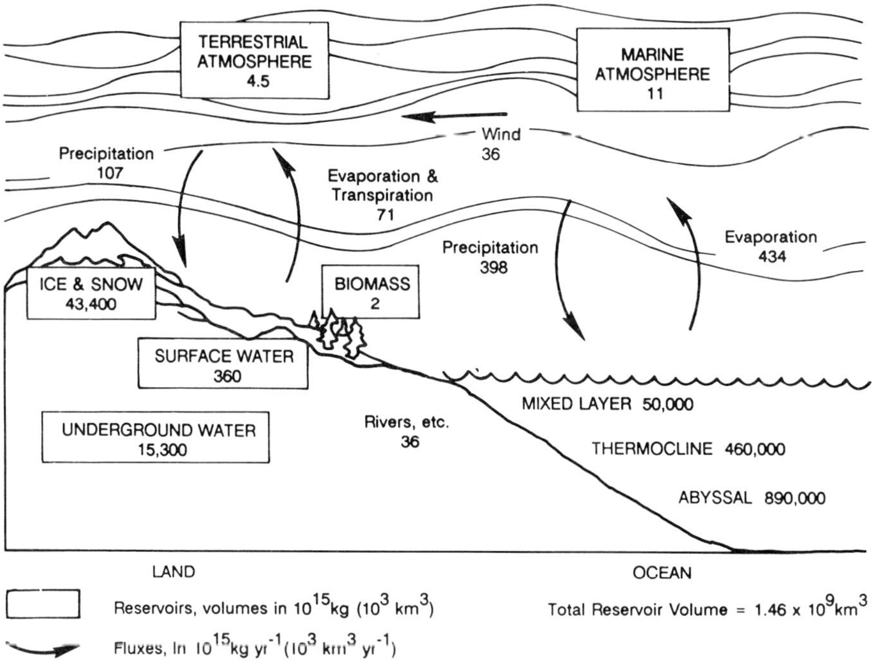

Fig. 5. The climatic hydrologic cycle at global scale (from NRC, 1986).

to the atmosphere by volcanism. Here, we call this the "tectonic hydrologic cycle." We illustrate it in Figure 6, but its characteristics are beyond the scope of this discussion. We focus here on those changes involving the earth's atmosphere and crust.

4. The biogeochemical basis of hydrologic science

Water has unique physical and chemical properties that enable it to play key roles in regulating the metabolism of our living planet:

Elixir of life — Water is practically a universal solvent. Because cell membranes are permeable only to dissolved substances, water is the elixir of life, essential both for the nourishment of cells and for the removal of their wastes. It plays this same role, both blood and lymph, at all higher levels of life's organization; the individual plant or animal, the household, city, civilization, and, apparently, for earth itself.

Climatic thermostat — Water's high specific heat gives it a large thermal inertia making it the flywheel of the global heat engine.

Global heat exchanger — Water's high latent heat makes it an efficient working fluid for the atmospheric heat engine converting solar energy into atmospheric and oceanic motions which act to redistribute both the energy and the water globally.

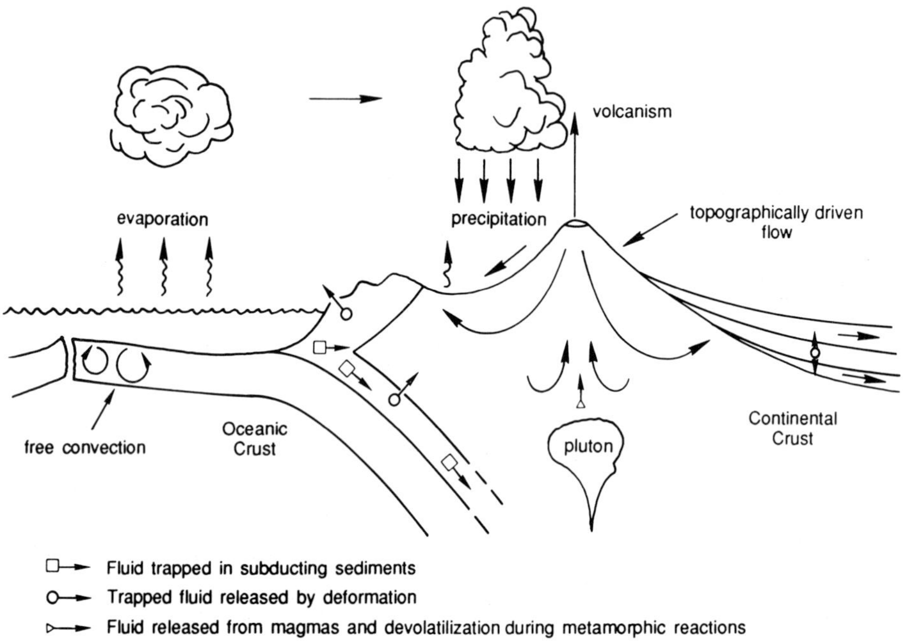

Fig. 6. The tectonic hydrologic cycle (from Foster and Smith, 1990).

Insulator of life — Water is less dense in its solid state than its liquid state making water bodies freeze from the top down. This maintains water temperatures conducive to many live forms.

These properties mean that water-soluble elements will follow the hydrologic cycles at least part way. If they are in a chemical compound that is volatile as well as soluble, the elements will follow the climatic hydrologic cycle all the way. The climatic hydrologic cycle is thus the integrating process for the fluxes of water, energy, and chemical elements. It is the fundamental biogeochemical cycle.

5. Some global change issues in hydrologic science

Because of the *interactive* nature of water in the Earth system, hydrology is both actively and passively important in global change. In the passive sense, the occurrence and quality of water are sensitive to atmospheric dynamics and composition. In the short term, the global distributions of rainfall, snowfall, evaporation and accumulated water, both surface and subsurface, affect the local extent and global distribution of biomass and biological productivity. Water movement couples the land with the oceans through the solution, entrainment, and transport of minerals and sediments; both liquid water and ice are powerful agents of erosion and, in

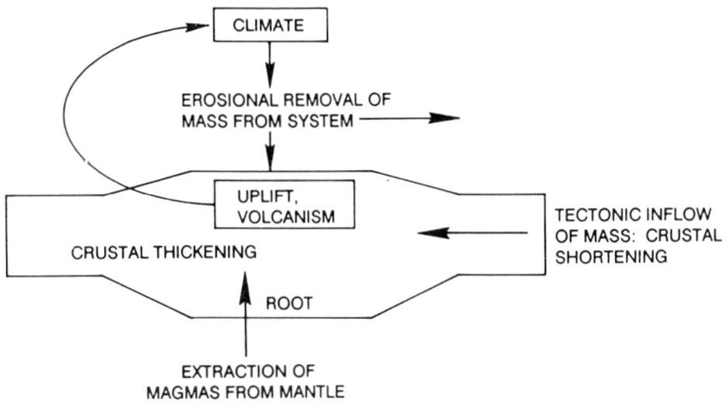

Fig. 7. The tectonic-climatic feedback loop (courtesy of B.L. Isacks and the Cornell Andes project).

the long term, they join with plate tectonics in shaping the land surface (Figure 7). In the active sense, changes in landsurface properties and vegetation cover may, among other things, alter the long-term average runoff of water which would demand a corresponding change in the long-term average convergence of atmospheric moisture. This, in turn, would require a commensurate adjustment in atmospheric dynamics.

The historical evolution of hydrologic science has been in the direction of ever-increasing space and time scale, from small catchment to large river basin to earth system, and from storm event to seasonal cycle to climatic trend. Inevitably, increased scale brings increased complexity and increased interaction with allied physical, chemical, biological, and mathematical sciences. New questions arise, many of which have implications for global change. We outline an eclectic subset of these here.

5.1. How do we aggregate the dynamics at various space–time scales in the presence of great heterogeneity?

This fundamental statistical-dynamical problem of hydrology remains unsolved and is crucial to the fidelity of atmospheric general circulation models (GCMs).

As far as global change is concerned, general circulation models are, for better or for worse, the only *prognostic* tool available. Furthermore, because of the complex interaction among the many physical, chemical, and biological components of the earth system, these models are a principal *diagnostic* tool in assessing system sensitivities. We must work to improve them.

The fluxes of moisture across the land surface boundary of GCMs depend upon the hydraulic conductivity (cm d^{-1}) of the soil. Such properties of natural soil formations are highly variable spatially as is seen by the observations plotted in Figure 8. In this illustration, Plots 1 and 2 are separated by only 50 m, and we see that their conductivities, for common soil water content, may differ by a factor of 100. With the fluxes being complex, nonlinear functions of conductivity and of the similarly variable water content, the GCM requirement for 10^4 km^2

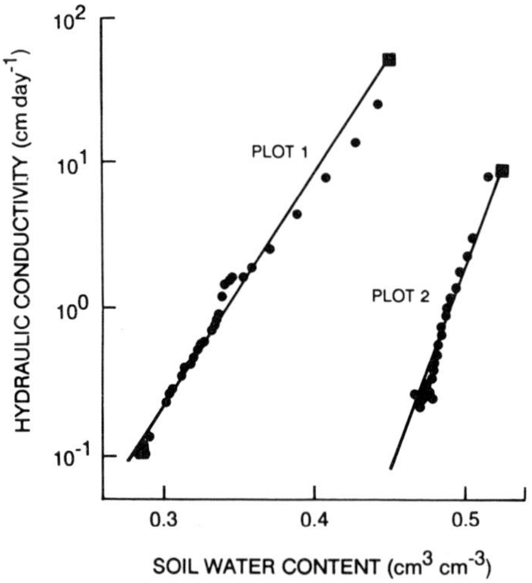

Fig. 8. Typical spatial variability of a natural field soil (from Nielsen et al., 1973).

gridsquare average fluxes is formidable. This problem occurs repeatedly in the parameterization of subgrid-scale processes for GCMs.

5.2. What are the feedback sensitivities of atmospheric dynamics and climate to changes in landsurface hydrology?

The fluxes of energy and water from landsurface to atmosphere are intimately linked and depend strongly upon the moisture state of the surface. The atmospheric processes responsible for supplying this moisture are themselves sensitive to the moisture-driven fluxes in a feedback process of variable importance to weather and climate depending upon such factors as latitude and season. While we can reason from first principles as to the sense and conditional magnitude of each of these feedbacks considered independently, their complex coupling demands that we estimate the net effect of landsurface change using a model that incorporates all the essential interactions. This work is in its infancy.

It seems logical to expect the sensitivity of atmospheric dynamics to landsurface change to be largest where the primary atmospheric motions are vertical as a result of moist convection (e.g. tropical latitudes). Similar reasoning suggests that this sensitivity should decline wherever lateral atmospheric motions and large-scale condensation dominate (e.g., middle latitudes).

Figure 9 illustrates the type of "experiment" needed. Here, using the GISS-II GCM is the simulated seasonal sensitivity of atmospheric moisture convergence to a drastic change in the soil moisture-holding capacity of the land surface, all else remaining constant. Note that the tropical climate (Figure 9b) exhibits sensitivity year-round, while in middle latitudes (Figure 9a) the sensitivity is confined to the growing season, at which time plants make deep soil moisture available to the

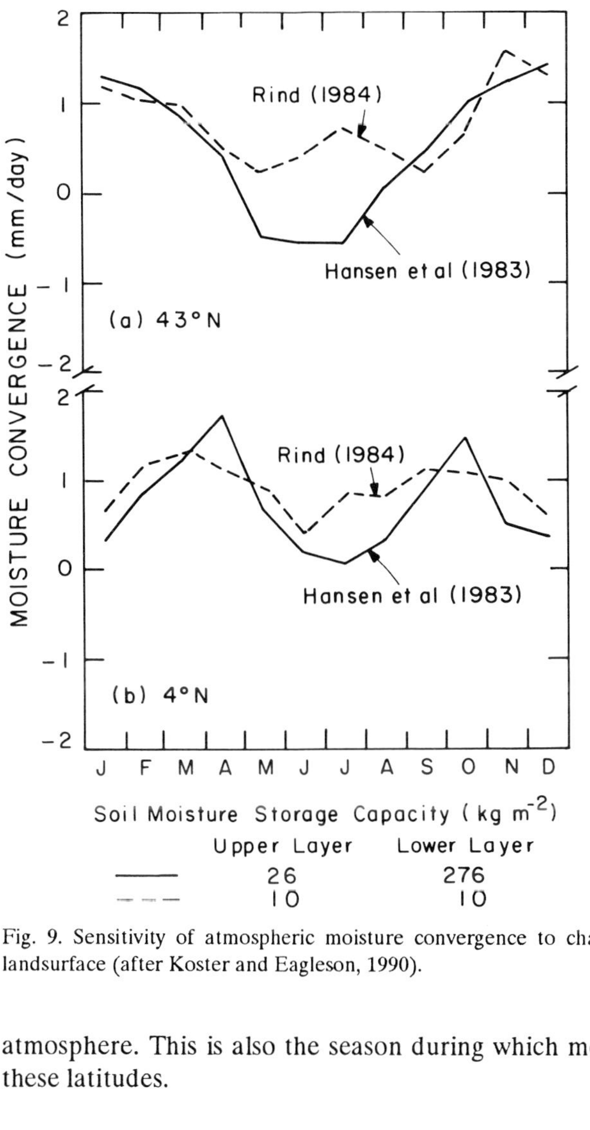

Fig. 9. Sensitivity of atmospheric moisture convergence to change in moisture-holding capacity of landsurface (after Koster and Eagleson, 1990).

atmosphere. This is also the season during which moist convection is important at these latitudes.

5.3. *What are the sensitivities of the methane productivity of wetlands to climate change?*

Methane gas follows water vapor and carbon dioxide in importance as a radiatively active "greenhouse" gas. Like carbon dioxide, its presence in the atmosphere is growing at well over 1% per year (see Figure 10). The major sources are thought to include wet ecosystems such as rice paddies and natural wetlands, as well as fossil deposits, combustion, and the digestive systems of cattle. In wetlands, methane is produced in saturated environments where it is readily released to the

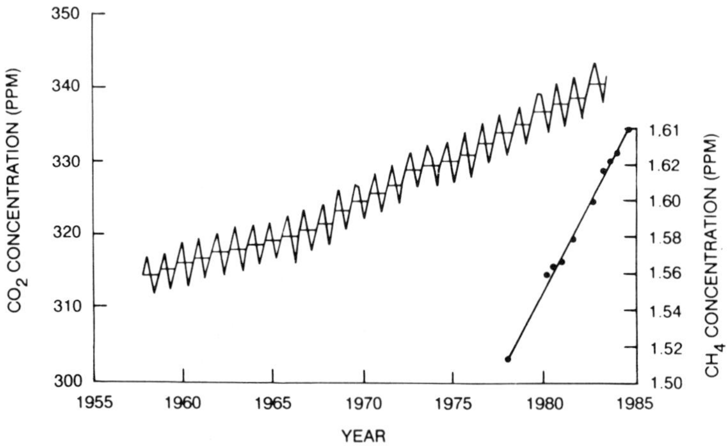

Fig. 10. Changing atmospheric concentration of carbon dioxide (CO_2) and methane (Ch_4) (from UCAR, 1985).

nearby atmosphere. Vegetation and waterlogged organic sediment together create the oxygen-deficient environment that encourages methane-producing microbes. The sediment also plays a regulatory role in release of the methane, in part by supporting methane-consuming microbes near the surface.

Hydrologic activity influences each step of this process. Water table elevation fixes the fraction of the organic sediment that is anoxic, and hence methane-producing, as opposed to the upper, unsaturated, methane-consuming sediment. To the extent that released methane may influence climate change, there is a potential for hydrologic feedback to the methane production.

5.4. *What is the physical basis for the observed geographical distribution of the major vegetation types on earth's continents?*

The vegetation cover of earth's continents is coupled physically to the climate system and to the soil by the fluxes of thermal energy, moisture, and nutrients. Different vegetation types have different physiological needs and tolerances which have developed over time in a process of co-evolution with their soil and climate partners. The current manifestation of this evolution and migration, as modified by human activity, is the observed distribution of vegetation formations such as is shown in Figure 11 for eastern North America. What are the physical reasons for a particular formation being confined to a certain geographical zone? It is important to gain quantitative understanding of these boundaries if the effects of climate change are to be accurately forecast.

The transitions from one vegetation type to another are known as ecotones. Because they represent marginal conditions, ecotones will be sensitive to changing climate, and predicting their location is an important test of the GCMs on which our climate forecasts are based.

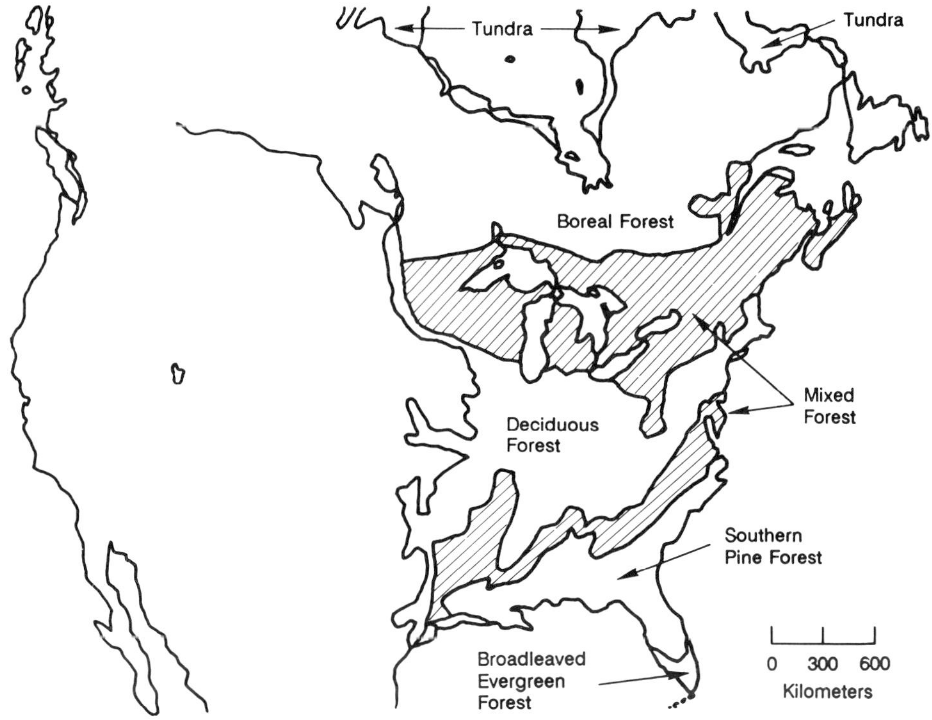

Fig. 11. Forest formations of eastern North America (map from Little, 1971; boundaries from Eyre, 1968; as presented by Arris and Eagleson, 1989).

Hydrology plays a role in the location of these ecotones. In some cases (e.g. savannas), water availability controls the configuration of the formation. In others, where water is plentiful, dominance in a given environment may go to the vegetation type that produces the most biomass, and hence wins the race to reproduce. Biomass production is a function of carbon assimilation through the process of photosynthesis. Since both carbon dioxide and water vapor exchange occur through the open stomata, the evapotranspiration rate is often used as the basis for estimating productivity. Such an estimate (Arris and Eagleson, 1989) apparently explains the transition between boreal and deciduous forest in eastern North America.

5.5. Can the dynamics of multistable nonlinear systems suggest new physical insights into the patterns of annual rainfall time series?

In now-classical work, E.N. Lorenz (1963) demonstrated that nonlinear phenomena can be the sources of intrinsically generated complex behavior. Although governed by well-defined deterministic equations, the dynamics may evolve into an aperiodic random-looking behavior called "deterministic chaos". A characteristic of such

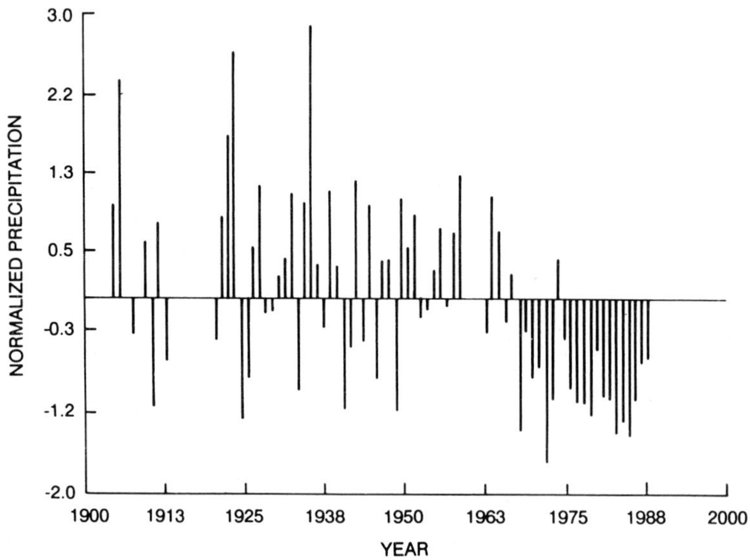

Fig. 12. Normalized rainfall departures, in standard deviations, at the Kaedi station (Mauritania) for the period 1904–1988 (from Demarée and Nicolis, 1990).

systems is a multiplicity of simultaneously available states called "attractors" among which the system chooses according to the initial conditions. Since the initial conditions may be subject to external perturbations, a multi-stable nonlinear system exhibits an aperiodic series of jumps between attractors.

Figure 12 shows the normalized precipitation time series for the western Sahel which shows an apparent abrupt transition from one state to another. This behavior can be modeled (see Demarée and Nicolis, 1990) as a bi-stable nonlinear dynamical system in which one stable state is quasi-normal rainfall and the other is low rainfall. The system is modeled as subjected to one stochastic perturbation such as would arise from the "normal" advective and convective processes and another of external origin such as sea surface temperature anomalies. The resulting behavior looks very much like Figure 12.

Such modeling exercises provide valuable new insights into hitherto perplexing observations. Estimating the model parameters from the data will, of course, allow potentially useful statistical predictions of such practical quantities as the residence times in the separate states.

These and many other fundamental problems of hydrology must be addressed in order to provide the ingredients for solving the sharpening conflicts of man and nature. Many, if not most, of them will require coordinated multidisciplinary field studies of the appropriate scale. Others, such as the measurement of unknown oceanic precipitation and evaporation, will require sensors, often satellite-borne, that are still undeveloped. Progress in many areas of the science is currently data-limited.

6. Hydrologic science is a distinct geoscience

This realization of the importance of water to the earth system at geophysical space and time scales has profound implications for the research and educational infrastructure and hydrologic science. We cannot build the necessary scientific understanding of hydrology at global scale from research and educational programs designed to serve the pragmatic needs of the engineering community. Hydrologic science must take its proper place as a geoscience alongside the atmospheric, oceanic, and solid earth sciences. A new definition of hydrologic science is needed and has been proposed recently (NRC, 1990) to reflect these evolving perceptions and to specify more clearly for programmatic purposes the current boundaries of the science.

First, the new perceptions: We add that hydrologic science is a geoscience interactive with the ocean, atmospheric, and (solid) earth sciences as well as with plant and animal ecologies and with human activities on a wide range of space/time scales. The new perceptions are the interaction of the components and the range of scales.

Second, the practical boundaries: The ubiquity of water on earth and its indispensability to life do not make hydrologic science out of all geoscience and biology. To form a separate identity we must specify and claim the central elements of our science and locate its administrative boundaries as an adjustable compromise between completeness and precedent. Here we disclaim any implied responsibility of hydrologic science for developing understanding of the physics, chemistry, and biology of water within the ocean and atmosphere reservoirs, for these processes lie firmly in the recognized domains of our sibling geosciences. We do include the study of snow, ice, and lakes within our definition of hydrologic science, however.

To establish and retain our individuality as a separate geoscience, we limit the scope of hydrologic science to the following:

Continental water processes — the physical and chemical processes characterizing or driven by the cycling of continental waters (solid, liquid, and vapor) at all scales (from the micro-processes of soil water to the global processes of hydroclimatology) as well as the biological processes which interact significantly with the water cycle.

(This restrictive treatment of biological processes is meant to include those which are an active part of the water cycle such as vegetal transpiration and many human activities, but to exclude those which merely respond to water, such as the life cycle of aquatic organisms.)

Global water balance — the spatial and temporal characteristics of the water balance (solid, liquid, and vapor) in all compartments of the global system: atmosphere, oceans, and continents.

(This includes water masses, residence times, interfacial fluxes, and pathways between the compartments. Other than for continents, including their aquifers, rivers, lakes, and glaciers, this does not include processes internal to the compartments.)

These boundaries of the science are illustrated schematically in Figure 13.

The complex problems of global change illuminate the multi-disciplinary nature of hydrologic science and make clear the need for extensive cross-discipline

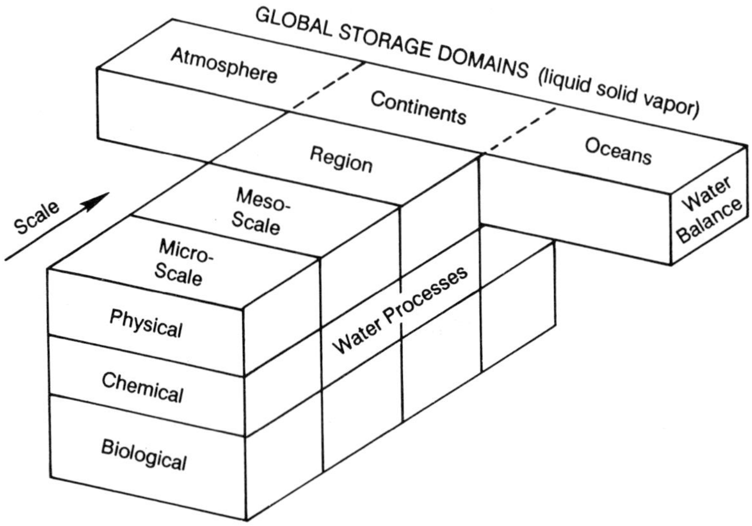

Fig. 13. Hydrologic science: a distinctive definition (from NRC, 1990).

interaction in education as well as in research. The problems pay no attention to organizational boundaries; thus, there are major areas of overlapping interest with other sciences and frequent needs to violate the stated boundaries. This is both inevitable and desirable. Primary in this regard is the case of landsurface–atmosphere interaction. Understanding the feedback effect of landsurface moisture and energy fluxes on the formation of weather and climate is vital; more traditional concerns for generalization of the space–time variability of storm precipitation demand incorporation of considerable atmospheric dynamics and thermodynamics. Similar trespassing must occur in the areas of fluvial geomorphology, micrometeorology, and plant ecology (to name but a few) due to the importance to the hydrologic cycle of such related processes as erosion, energy flux, and transpiration.

7. Needed actions

Development of hydrology as a science is important to the current effort to understand the interactive behavior of the earth system because of the key role that the hydrologic cycles is now known to play therein. Achieving this comprehensive understanding will require the kind of long-term disciplinary and multidisciplinary effort that can only be sustained by a vigorous scientific infrastructure. To advance the science of hydrology resources will be needed:

– Separate research grant programs should be established in hydrologic science to help create and maintain a cadre of hydrologic scientists and to provide a focused image and identity for the science.

– Coordinated, multi-disciplinary field experiments should be increased in number particularly at the mesoscale and in a variety of climates.

– Long-term hydrologic observations should be given renewed attention.

– Graduate education in hydrologic science should be pursued independently of engineering in new programs based within the geosciences.

Acknowledgements

This material has been drawn in large part from the collective work of the NRC Committee on Opportunities in the Hydrologic Sciences (P.S. Eagleson, Chairman, W.H. Brutsaert, S.C. Colbeck, K.W. Cummins, J. Dozier, T. Dunne, J.M. Edmond, V.K. Gupta, G.C. Jacoby, S. Manabe, S.E. Nicholson, D.R. Nielsen, I. Rodríguez-Iturbe, J. Rubin, J.L. Smith, G. Sposito, W.T. Swank, E.J. Zipser), including contributions from outside the committee. Of the latter, particular credit is given here to H.F. Hemond for the material on methane, and to C. Nicolis for that on chaos. The opinions expressed are those of the author, however, and do not necessarily reflect those of the committee or of the NRC.

The report of the committee is entitled "Opportunities in the Hydrologic Sciences" and is available from the National Research Council, National Academy Press, 2101 Constitution Avenue, N.W., Washington, D.C., 20418, ISBN 0-309-04244-5.

References

Ad Hoc Panel on Hydrology, 1962. Scientific Hydrology. U.S. Federal Council for Science and Technology, Washington, D.C., 37 pp.

Arris, L.L. and Eagleson, P.S., 1989. A physiological explanation for vegetation ecotones in eastern North America. MIT Dep. Civil Eng., R.M. Parsons Lab. Rep., No. 323, 247 pp.

Demarée, G.R. and Nicolis, C., 1990. Onset of Sahelian drought viewed as a fluctuation induced transition. Q. J. R. Meteorol. Soc. (in press).

Entekhabi, D., 1989. Landsurface hydrology parameterization for atmospheric general circulation models: inclusion of subgrid scale spatial variability and screening with a simple climate model. Ph.D. thesis, Dep. Civil Eng., MIT, Cambridge, Mass.

Eyre, S.R., 1968. Vegetation and Soils: A World Picture. 2nd edition, Arnold Press, 327 pp.

Forster, C. and Smith, L., 1990. Fluid flow in tectonic regimes. In: Fluids in Tectonically Active Regimes of the Continental Crust. Mineral. Assoc. Can. (in press).

Hansen, J., Russell, G., Rind, D., Stone, P., Lacis, A., Lebedeff, S., Ruedy, R. and Travis, L., 1983. Efficient three-dimensional global models for climate studies: Models I and II. Mon. Weather Rev., 111: 609–662.

Horton, R.E., 1931. The field, scope, and status of the science of hydrology. Trans. AGU, Reports and Papers, Hydrology, National Research Council, Washington, D.C., pp. 189–202.

Koster, R.D. and Eagleson, P.S., 1990. A one-dimensional interactive soil–atmosphere model for testing formulations of surface hydrology. J. Climatol., 3: 593–606.

Little, E.L., 1971. Atlas of United States Trees: Vol. 1, Conifers and Important Hardwoods. USDA Misc. Publ. 1141, GPO, Washington, D.C.

Lorenz, E.N., 1963. Deterministic nonperiodic flow. J. Atmos. Sci., 20: 131–141.
Meinzer, O.E. (Editor), 1942. Hydrology, Physics of the Earth — IX. McGraw-Hill, New York, N.Y. (republished by Dover Publications, New York, N.Y.).
National Research Council, 1982. Scientific Basis of Water-Resource Management. National Academies Press, Washington, D.C., 127 pp.
National Research Council, 1986. Global Change in the Geosphere–Biosphere. National Academies Press, Washington, D.C., 91 pp.
National Research Council, 1990. Opportunities in the Hydrologic Sciences. National Academies Press, Washington, D.C.
Nielsen, D.R., Biggar, J.W. and Erth, K.T., 1973. Spatial variability of field measured soil water properties. Hilgardia, 42: 215–259.
Rind, D., 1984. The influence of vegetation on the hydrologic cycle in a global climate model. In: J.E. Hansen and T. Takahashi (Editors), Climate Processes and Climate Sensitivity. Geophysical Monograph 29, Maurice Ewing Vol. 5, AGU, Washington, D.C., pp. 73–91.
UCAR, 1985. Opportunities for Research at the Atmosphere/Biosphere Interface, Universities Corporation for Atmospheric Research, Boulder, Colo.

Chapter 16

The theory of the hydrologic model, or: the struggle for the soul of hydrology

M.B. ABBOTT

Toutes choses sont dites déjà; mais comme personne n'écoute, il faut toujours recommencer.
— André Gide

Abstract

After an introduction to the basic problem of maintaining continuity of thinking through our current discontinuity in thought, the function of the sign within the process of theorising is considered. As understanding then proceeds through the ordering of signs, the two basic levels of orderings, of *mathesis* and *taxinomia*, are considered in relation to hydrologic theory. It is shown that hydrology cannot itself be a science, even though it draws upon many sciences, and it cannot itself be a technology, even though it serves many technologies. The nature of hydrology is identified, and with this the origin of its current theoretical despair. The means of overcoming this despair are then intimated.

1. Introduction: maintaining continuity in thinking through a discontinuity in thought

Hydrology is like the numbers: it does not exist (e.g. Cooke and Bez, 1984, p. 3). Since, however, such ontological details have never yet inhibited the flights of scientific theorising, we must suppose that a theory of hydrology has every bit as much right to exist as does a theory of numbers. Perhaps it is with hydrology as it was with language in the Classical age according to Foucault (1966, p. 93/79[1]), that *in the limit* one could say that, although this language no longer existed in the original sense of language, as a direct translation of thought, it continued to *function*, in that it continued to appear in the form of various *roles*, by way of *representation*: in the limit it came to refer only to itself (see also Heidegger, 1927/1962, p. 168/212). When looking at the matter in this way, we remark a resemblance to the status of the angels described by Barth (1950/1960, e.g. p. 484), in that,

[1] The first page numbers refer to the original work and the second to the English translation given in reference. Any reader who senses in any way at all the relevance of the present contribution must return to consult these references, and then preferably in their originals, for it is only thereby that the material introduced here will come fully into focus.

Biblically, the angels compensate for their *ontological weakness* precisely by the strength of their *functional reality*. Indeed, if we may be permitted to compare the status of hydrology to that of angelology, we must suppose, if only on historical grounds alone, that such a nebulosity in existence provides the most fertile ground imaginable for theorising (e.g. Barth 1950/1960, pp. 377, 378).

Thus there is every reason to believe that hydrology has a theory and thereby that there exists a theoretical hydrology. (See, more generally, Eco, 1977, e.g. p. 158.)

We live however, for better or for worse, in interesting times, where *change* often takes on the main attributes of a blast wave. Change then has the effect that the familiar runs of our mental burrows become repeatedly dislocated, at best rendered uninhabitable and as often as not blocked completely. We thereby find ourselves forever struggling to find new ways between sustenance and shelter in our mental worlds.

Now it is exceedingly distressing to thinking of any theoretical inclination to become the subject of such essentially discontinuous changes in the nature of the thinking process itself: all at once the familiar relations and functional forms disappear, so that the established limits suddenly become irrelevant. In any such situation, it is no bad idea for us *to take stock* of the changed configuration of our mental world before we take our own actions. This is to say, however, that all at once, and in the first place, we have to *talk very much less* so that we may *listen very much more*. (And that must of course be well-nigh impossible for a politician!)

But to whom and to what should we listen? To the voice of history? But then what kind of historiography is suited to treating a discontinuity in thought, to enunciating a history of discontinuity in thought? There is a problem here already: we have to use the thought world that we do have in order to represent another world which was thinking or is coming to think in a very different way. This is a central problem of hydrology, just as it is of almost every other theoretical endeavour of our time.

2. Navigating through a world of models: the function of the sign

So far we have been able to navigate through life, *qua* individual and *qua* society, by constructing models of the world. The problem is that we have now increasingly to navigate through a world of models. In effect, we have now to *navigate through machine-encapsulated knowledge* (Guay, 1990). This is just as much an imperative for the hydrologist as it is for any other professional person.

But what are these models that populate our new knowledge worlds? Let us put aside our usual conceptions and the pet definitions that we are accustomed to expounding. Let us consider the model in *its* functional reality. In what way does it function? Nowadays most of us would not hesitate long before saying that it functions as a piece of equipment, or as a *tool*. Indeed, referring to the current informational-technological literature we see that we now have tool sets, tool boxes, tool benches and you name it in tools! But although tools have always been our means of accomplishing something, we must still return to ask here: *how* is it that these tools now *function.* We can perhaps best listen first to Heidegger (1927/1962, pp. 77, 79, 80/107, 108, 110, 111):

> We come across equipment in signs. The word 'sign' designates many kinds of things; not only may it stand for different *kinds* of signs, but Being-a-sign-for can itself be formalised as a universal kind of relation, so that the sign-structure provides an ontological clue for 'characterising' any entity whatsoever.
>
> But signs, in the first instance, are themselves items of equipment whose specific character as equipment consists of *showing* or *indicating.*
>
> ... Signs ... let what is ready-to-hand be encountered; more precisely, they let some context of it become accessible in such a way that our concernful dealings take on an orientation and hold it secure. A sign is not a Thing which stands to another Thing in the relationship of indicating; it is rather *an item of equipment which explicitly raises a totality of equipment into our circumspection so that together with it the worldly character of the ready-to-hand announces itself.* In a symptom or a warning-signal, 'what is coming' 'indicates itself', but not in the sense of something merely occurring, which comes as an addition to what is already present-at-hand; 'what is coming' is the sort of thing which we are ready for, or which we 'were not ready for' if we have been attending to something else. In signs of something that has happened already, what has come to pass and run its course becomes circumspectively accessible. A sign to mark something indicates what one is 'at' at any time. Signs always indicate primarily 'wherein' one lives, where ones's concern dwells, what sort of involvement there is with something.
>
> The peculiar character of signs as equipment becomes especially clear in 'establishing a sign' ["Zeichenstiftung"]. This activity is performed in a circumspective fore-sight ["Vorsicht"] out of which it arises, and which requires that it be possible for one's particular environment to announce itself for circumspection at any time by means of something ready-to-hand, and that this possibility should itself be ready-to-hand. But the Being of what is most closely ready-to-hand within-the-world possesses the character of holding-itself-in and not emerging. Accordingly our circumspective dealings in the environment require some equipment ready-to-hand which in its character as equipment takes over the 'work' of *letting* something ready-to-hand *become conspicuous.* So when such equipment (signs) gets produced, its conspicuousness must be kept in mind. But even when signs are thus conspicuous, one does not let them be present-at-hand at random; they get 'set up' ["angebracht"] in a definite way with a view towards easy accessibility.

So all signs serve as equipment, while nowadays a model is clearly itself formalised entirely in signs. In our age of the programmable machine it becomes more evident than ever before that *a model is a collection of signs that serves as a sign.* The change in the nature of our thinking existences — and this is not restricted to our professional existences — is that we must now pass through a discontinuity in our use and experience of signs.

Now there is nothing new about *this* kind of experience: history has its usual stock of precedents and precepts. We can then best listen next to Foucault (1966, p. 72/58) reconstructing the consequences of one of the last such shock waves to strike our civilization[2]:

> What is the sign in the Classical age? For what was altered in the first half of the seventeenth century, and for a long time to come — perhaps right up to our own day — was the entire organization of signs, the conditions under which they exercise their strange function; it is this, among so many other things one knows or sees, that causes them to emerge suddenly as signs; it is their very being. On the threshold of the classical age, the sign ceases to be a form of the world; and it ceases to be bound to what it marks by the solid and secret bonds of resemblance or affinity.

Rather as, in the first half of the 17th century, language became much less a direct reflection of thought, coming in a certain limit no longer to refer directly to elements of thought at all, but only to itself, as in the various jargons of the specialities (law, medicine, botany, mathematics, and especially algebra, etc.), so today we observe a further stage in this process, where, in another kind of limit, our very thinking existence becomes only a representation or reflection of language. Today, in this other kind of limit, it is increasingly the television screen and its Siamese twin, the computer screen, which establish our mental realities, and even our very sense of reality (*sens de réel*). In this limit, we already live in a world of electronically-encapsulated "virtual realities".

---

[2] Foucault, together with Derrida, represent in philosophy a current of contemporary thought that is generally called *post-modernism*. Its most eloquent expressions are nowadays to be found all around us, in architecture. This current of thinking is characterized, however, precisely by its insistence on *listening*, and specifically on listening to the "view of the past". Modernism was so full of itself that it proclaimed only itself and deliberately shut-off its awareness of the past. Every new artefact, even a glass-box building, was thus seen as inherently superior to anything that it replaced and so to anything that it came to accompany. Post-modernism opposes modernism essentially over this relation of our spatial experiences to time. Modernism "lives with the times", post-modernism tries more to live "out of time".

From all that we have learned, and at least since Heidegger (1927/1962), we must realize however that every such attitude taken towards time is also and inseparably an attitude taken towards Being: *Sein und Zeit*, Being and Time. Thus modernism cultivates and emphasizes one manner of our Being-in-the-world (Heidegger's *In-der-Welt-sein*), while post-modernism inculcates a quite other manner of essential existence. [Concerning the influence of Foucault and Derrida in modern technology, see further Fraser Taylor (1991, p. 10), Harley (1989), and Abbott (1991a, p. 9). More generally, and also as a background to the brothers Dreyfus' critique of the symbolic paradigm of Artificial Intelligence (AI), see Dreyfus and Rabinow (1989).]

In their delightful sketch of a visit to a fictive STARDUST software house, Andersen and Mathiessen (1987, p. 409) give the following words to the singularly pragmatic head of STARDUST's production department:

> Humanities do not look upon language as a reflection of something else, but as man's link to Nature and fellow human beings. Man creates language through his association with the world, and at the same time language determines how we associate with and understand this world. This is also true where communication executed through programs is concerned. Programs create the reality experienced by users of computers. Remember that the computer is a tool for simulation, and what is simulated becomes the reality of the user. In a society like ours — the post-modern society — there are no 'great stories' to justify a specific perception of reality like there were in the 19th century. We should rather see the situation thus: communication is based on a number of language games which are played according to specific sets of rules. Each group of society can 'play a game', and it is the efficiency of each game that justifies it. The computer medium should be seen as a technical device which allows its owner to play particularly efficient games. The execution of the program must serve an intention with the sender (and to do that it must also often serve the receiver). Whether the execution is good or poor depends in reality on how well it functions. That is, how close it comes to a specific objective. A good program is one that creates the reality intended by the sender in the most efficient way.

3. Understanding through the ordering of signs

At such a discontinuity in the history of thought we naturally also encounter a discontinuity in "what it means to understand" (see, within the ambit of Computational Fluid Dynamics, Chorin, 1983). At one time something is understood on the basis of one body of knowledge, while only a short time later this same body of knowledge provides no understanding at all, but instead a quite other kind of knowledge is required before understanding is experienced. Continuing our search for some variable that remains continuous, however, we follow Foucault when he equates understanding with *ordering*. Thus (1966, pp. 86, 87/72):

> What makes the totality of the Classical *episteme* possible is primarily the relation to a knowledge of order. When dealing with the ordering of simple natures, one has recourse to *a mathesis*, of which the universal method is algebra. When dealing with the ordering of complex natures (representations in general, as they are given in experience), one has to constitute a *taxinomia*, and to do that one has to establish a system of signs. These signs are to the order of composite natures what algebra is to the order of simple natures. But in so far as empirical representations must be analysable into simple natures, it is clear that the taxonomy relates wholly to the mathesis; on the other hand,

> since the perception of proofs is only one particular case of representation in general, one can equally well say that mathesis in only one particular case of taxonomy. Similarly, the signs established by thought itself constitute, as it were, an algebra of complex representations; and algebra, inversely, is a method of providing simple natures with signs and of operating upon those signs.
>
> ...At the two extremities of the Classical *episteme*, we have a *mathesis* as the science of calculable order and a *genesis* as the analysis of the constitution of orders on the basis of empirical series. On the one hand, we have a utilization of the symbols of possible operations upon identities and differences; on the other, we have an analysis of the marks progressively imprinted in the mind by the resemblances between things and the retrospective action of imagination. Between the *mathesis* and the *genesis* there extends the region of signs — of signs that span the whole domain of empirical representation, but never extend beyond it. Hedged in by calculus and genesis, we have the area of the *table*. This kind of knowledge involves the allotting of a sign to all that our representation can present us with.

In effect, the first half of the 17th century initiated new ways to describe the nature of structure so that it could better understand the structure of nature. We, now, in our own time, have to perform a rather similar trick![3]

Every model presupposes at least one language, which is the language which is common to the producer of the model and its user. This is the language of the *universe of discourse* of the model (e.g. Genesereth and Nilsson, 1988, pp. 10, 15, 19, 22). Models are accordingly classified in the first place by this primary language which they employ. Thus, for example, when we speak of a "physics-based, spatially-distributed hydrologic model", we refer at once to a rather comprehensive language which we may expect to be common to all users and all producers of such models. Every language has, in its turn, a *grammar* and a *discourse*, and these in their turn presuppose a *rhetoric*. Let us, however, be very careful about how we understand and use the notions of "grammar" and "discourse" for, as we shall see, this understanding is essential to our further understanding of the nature of hydrologic theory. As a result of his own listening to Locke and Hobbes, Foucault (pp. 96–98/82–84) expressed his understanding of these notions in terms of ordinality as follows:

---

[3] There might be a temptation nowadays just to equate *mathesis* with mathematics and *taxinomia* with taxonomy. This would, however, be misleading given the current standard, modernist — and therefore very restricted and one-sided — interpretations of mathematics and, albeit to a lesser degree, taxonomy. If we consider instead the original Greek context of these terms, which is the context of a *game*, then *mathesis* has to do with the decision to play the game and the gathering together of the wherewithal to do so, *taxinomia* refers to the appearance of "names", or the "giving of names", which process defines the rules of the game, and *genèse*, the "genesis", appertains to the playing-out of the game in practice.

What distinguishes language from all other signs and enables it to play a decisive role in representation is, therefore, not so much that it is individual or collective, natural or arbitrary, but that it analyses representation according to a necessarily successive order: the sounds, in fact, can be articulated only one by one; language cannot represent thought, instantly, in its totality; it is bound to arrange it, part by part, in a linear order. Now, such an order is foreign to representation.

...It is to thought and to signs what algebra is to geometry: it replaces the simultaneous comparison of parts (or magnitudes) with an order whose degrees must be traversed one after the other. It is in this strict sense that language is an *analysis* of thought: not a simple patterning, but a profound establishment of order in space. It is here that we find that new epistemological domain that the Classical age called 'general grammar'. It would be nonsense to see this purely and simply as the application of a logic to the theory of language. But it would be equally nonsensical to attempt to interpret it as a sort of prefiguration of a linguistics. *General grammar is the study of verbal order in its relation to the simultaneity that it is its task to represent.* Its proper object is therefore neither thought nor any individual language, but *discourse*, understood as a sequence of verbal signs. This sequence is artificial in relation to the simultaneity of representations, and in so far as this is so language must be in opposition to thought, as what is reflected upon is to what is immediate.

...In relation to the evident, necessary, universal order introduced into representation by science, and by algebra in particular, language is spontaneous and un-thought-out; it is, as it were, natural. It is equally, according to the point of view from which one looks at it, an already analyzed representation and a reflection in the primitive state. In fact, it is the concrete link between representation and reflection.

Language is the original form of all reflection, the primary theme of any critique. It is this ambiguous thing, as broad as knowledge, yet always interior to representation, that *general grammar* takes as its object.

But a certain number of consequences must at once be drawn here.

1. The first is that it is easy to see how the sciences of language are divided up in the Classical period: on the one hand, rhetoric, which deals with *figures* and *tropes*, that is, with the manner in which language is spatialized in verbal signs; on the other, grammar, which deals with articulation and order, that is, with the manner in which the analysis of representation is arranged in accordance with a sequential series. Rhetoric defines the spatiality of representation as it comes into being with language; grammar defines in the case of each individual language the order that distributes that spatiality in time. This is why, as we shall see, grammar presupposes languages, even the most primitive and spontaneous ones, to be rhetorical in nature.

2. On the other hand, grammar, as reflection upon language in general, expresses the relation maintained by the latter with universality. This relation can take two forms, according to whether one takes into consideration the possibility of a *universal language* or that of a *universal discourse.*

4. Hydrological rhetoric and its ordering in grammar

Having now listened for so long — and to a post-modernist French philosopher at that! — it is time for us to talk again for ourselves, if not as hydrologists, then at least about hydrology. The hydrological model is nowadays almost invariably programmed on a digital machine, and increasingly it is programmed using a variety of formal languages. Since, however, the programmable machine is itself purely a machine for manipulating signs, hydrologic information and knowledge become in this way mapped into a system of signs, ordered according to a certain, definite logic. The signs themselves can be grouped into programme entities that form a hierarchy according to our ability to map these entities into elements of our original universe of discourse of the subject area. Thus, for example, it will require very little effort on our part to select, or "pick up" a tree icon displayed by a Geographic Information system (GIS) and "drag" it to a defined area, so as to represent forested land with all of its attributes. Both the grammar and the discourse are exceedingly transparent in this case. Most of us would not have much more difficulty if we had instead to allocate an object class called "trees" to a particular set of coordinates, to the same effect. We would many of us have more difficulty in writing a code that actually allocated "trees" to an arbitrarily-chosen area, and then in such a way as to interrogate the user about their attributes, preferably in a natural language. If we were to proceed further again, below this level, down to the basic enabling operations of the compiler and the more general operating system, we should most of us nowadays be very lost indeed.

Thus we most conveniently allocate or order signs in hierarchies according to the ease of our own understanding. Each ensemble of such hierarchies that is relatively (more-or-less) independent of other hierarchies constitutes the matter of one or another particular science. Arriving at these uppermost levels, of the signs that have the greatest significance to us, we have a collection of statements about the matters of the particular sciences that are brought into play in our model, so that above these again we must have a "metalevel" which talks about the *relations* between these matters of the sciences which are so brought into play. The corresponding "metatheory" has as its central theme a language which gives "*the original form of all reflection, the primary theme of any critique*" in this region. It is the theory of this language which is once again subsumed under the one term of general grammar, of *Grammaire générale.*

Two examples are now in order to illustrate this thesis in the case of hydrology, as follows:

(1) Hydrology clearly provides a means of discourse, and often *the only* means of discourse between subjects, such us plant physiology and hydrogeology, which otherwise would have precious little to say to one another. Wherever we look, we see hydrology fulfilling this function of maintaining a discourse, or "interfacing" between otherwise rather separate and distinct, but none-the-less well-established sciences.

(2) Even when hydrology does come across some matters which it believes to be well and truly its own, it inevitably discovers sooner or later that these matters have been developed much more generally and at greater depth in other areas, and then again as integral parts of already-well-established sciences. A classical example is provided by "the theory of the unit hydrograph" and all of its various elaborations, which Dooge (1973) showed was little more than a partial reformulation of a general linear systems theory. Nowadays we subsume all such methodologies under the theory of generalised multiplication, as in the theory surrounding the convolution integral, such as is expounded in the many books on operational calculus, like those treating the Fourier transforms.

However, as we have just seen with Foucault, every grammar presupposes a rhetoric, which is that which fills out the space of the universe of discourse with (iconical and verbal, including mathematical) signs. The rhetoric has then again to fill out the spaces of *mathesis* according to *taxinomia*, as well as the coming to presence of these in their orderliness, their *genèse* (Foucault, 1966, p. 88/74). These two steps along our ways of ordering the world, and their coming to presence, can be followed quite clearly in the case of the physics-based, spatially-distributed modelling systems, such as the SHE (Abbott et al., 1985). Of the *mathesis* it should not be necessary to say very much: the equations that we collect together, such as those of Darcy and Richards, are familiar enough, as are the fun and games of constructing numerical representations of these. But hydrology is deeply involved also in *taxinomia*: the "where" of "where the game is played", the "how" of "how the rules are set". Studies involving geological substructures, profiles of soil types, crop water requirements, root zone properties and many others are essentially studies in *taxinomia* (e.g. Soil Conservation Service, 1975). It is these studies which tell us which rules to apply: which relations, which functions, which constants, and so on, have to be introduced in the associated parts of the model. This is to say that a theoretical hydrology has to do, in the first instance, with the setting out and ordering, in these two basic senses, of all information and all knowledge about water in its land phase.

We need, however, to relate our earlier theme of the connecting role of hydrology to its more general *taxinomia*. This connection arises from the fact that both the *mathesis* and the *taxinomia* of hydrology as so far treated are those of distinct and more-or-less exact sciences: hydrogeology, soil physics, agronomy and plant physiology in the above examples. This is to say that hydrology has a *taxinomia* the objects of which are themselves more-or-less complete and often disparate sciences. The principal business of hydrology is then one of *maintaining a discourse* between a number of different and otherwise rather separate sciences. In this respect, hydrology has to have a dimension of *listening* to the utterances of other sciences, as a basis for all of its own pronouncements. The central theory (and indeed the central dogma) of hydrology is thereby composed from the rhetorical and grammatical aspects of the language which it employs as it goes about this business.[4]

When seen in this light, some of the basic difficulties currently encountered in hydrology become more transparent. Probably no one has explicated these difficulties so trenchantly as Klemes, who has complained bitterly about the *misappropriation* of techniques borrowed from specific, and especially mathematical sciences (Klemes, 1986):

> Hydrology, having no solid foundation of its own and moving clumsily along on an assortment of crutches borrowed from different disciplines, has always been an easy victim of this practice. Every mathematical tool has left behind a legacy of misconceptions invariably heralded as scientific breakthroughs. The Fourier analysis, as was pointed out by Yevjevich (in 1968), had seduced the older generation of hydrologists into decomposing hydrologic records into innumerable harmonics in the vain hope that their reconstruction would facilitate prediction of future hydrologic fluctuations (fortunately few computers were available at the time so that the Fourier fever did not become an epidemic); various statistical methods developed for evaluation of differences in repeatable experiments have been misused to create an illusion of a scientific analysis of unrepeatable hydrologic events; linear algebra has served to transform the idea of a unit hydrograph from a crude but useful approximation of a soundly based concept into a pretentious masquerade of spurious rigor now exercised in the modelling of flood events; time series analysis has been used to remake inadequate 20-year streamflow records into 'adequate' 1000-year records, or even more adequate 10,000-year records, and the theory of pattern recognition is now being courted in the vain hope that it will lend scientific legitimacy to the unscientific concept of mindless fitting that dominates contemporary hydrologic modelling. In all these cases, mathematics has been used to redefine a hydrologic problem rather than to solve it.

From what has now been said, and after what has now been heard, we cannot be surprised any more about this dismal view of hydrology. The *cri de coeur* that we hear is a cry of despair, a despair that hydrology is not the science or even the technology that would place it on a par with the sciences upon which it draws or the technologies which it serves. But this state of affairs is in point of fact *essential* to hydrology. No theory of science — let alone a theory of sciences — can itself be constituted as a scientific theory. Similarly, that which serves a technology need not

[4] In the *Grammaire générale* we introduce the loose equivalences of *mathesis* — rhetoric, *taxinomia* — grammar and *genèse* — dialectic. Here we are dealing with just one realisation (instantiation) of this.

necessarily be a technology — and indeed it is most commonly not a technology. Hydrology, quite consentaneously, is not itself a science, and neither is it itself a technology, even though it deals in the productions of many sciences and serves many technologies. Accordingly, when viewed scientifically or technologically, it appears to be bereft of a body of theory and so lacking all intentionality: it appears to be without an own form and without an own direction. But if it is not a science and not a technology, what then *is* the status of hydrology? It is now the time and the place to state, provocatively but with serious intent, that *hydrology is a rhetoric waiting for a grammar.*

5. The struggle for the soul of hydrology

What is then being said here is that hydrology already in fact fills out a space of discourse, but it has not yet found any definitive means to order the elements contained within this space. These definitive means, which it still lacks, constitute, of course, its theory. But what is further implied, however, is that it has not yet found these means, its theory, precisely because it has so far been looking in the wrong place. By trying to constitute itself as a science, or even as a technology, it is trying to become something which it is not and never can be — and it is from this that its despair originates. In the technical terms introduced by Kierkegaard (1849), this despair is of the "α" or "feminine" kind: a despair that hydrology, in its despair, does not will to be itself, so that it is a despair in weakness. Despair, for Kierkegaard, was the outer form or symptom of a sickness of the spirit, the Biblical (John, XI: 4) "sickness unto death". It thus stands, in theological terms, in a particularly critical relation to the soul. To the extent that this despair turns inwards upon itself, so the soul stands to be lost; while to the extent that it is transcended, by faith, hope and charity, as the instruments of love as expressed in solicitude, so the soul stands to be redeemed. Hydrology may turn in upon itself, in a despair that is certainly driven by an awareness of a profound truth; or hydrology may transcend this despair, so as to become that "beautiful study" which it has always had the potential to become (Abbott, 1992). But it is the very mark of the significance of hydrology that it must under all circumstances meet with despair, and we owe much to Klemes for expressing the despair of hydrology so clearly, eloquently and convincingly.

Now, it follows from this analysis that hydrology cannot be alone in these travails: it must share these difficulties and problems, as they mount up to despair, with other subjects of a similar constitution. We can at once identify two of these subjects as *geography* and *ecology*. Can hydrology not learn something from their experiences; can they not in their turn learn something from the experiences of hydrology?

Geography has long had a grammar, which is the grammar of the *map*. The theory of the map is encompassed by *cartography*. How has our current change in thinking influenced geography through changing the system of ordering of signs that is the map, as directed by cartography (or perhaps in despite of direction from cartography)? Let us listen to Fraser Taylor speaking on "New directions for cartography in the information era" (1991):[5]

> It can be argued that the 1990s will be the 'Digital Decade' and that the already substantial impact of digital technologies on cartography will continue to accelerate. Despite the importance of issues such as data capture, data base structure, software, hardware and standards, the central issues which should determine new directions for cartography are not primarily technical. Indeed, it can be argued that if the prevailing technological paradigm is allowed to predominate then much will be lost. To allow our understanding and definition of cartography to be determined by a technological imperative would be a mistake.
>
> In 1984 the term 'A New Cartography' was introduced (Taylor, 1985) to draw attention to the dramatic impact of information technologies on the discipline. Part of the concluding sentence of the argument made in 1985 read, 'The greatest ... challenge facing cartography does not lie in terms of learning new techniques, but in creating a radically new concept for our discipline' (Taylor 1985, p. 22).
>
> It can be argued that this challenge remains today and, if anything, is more pressing. Ramirez has argued that, 'Unfortunately, the advent of new technologies such as computer-aided mapping, land information and geographic information systems (before the theoretical understanding of cartography was completed) decreased the interest and amount of research in theoretical cartography. As a result of that, there is no widely accepted cartographic theory yet that can be used as the basis for the cartographic process' (Ramirez, 1990, p. 19). The conceptual and theoretical development of cartography as a discipline has been retarded, if not diverted, by the explosion of interest in automated mapping and GIS. The neo-formalist and neo-positivist paradigms which have emerged as a result are somewhat sterile and too many modern cartographers are primarily technological specialists with a limited understanding of the problems to which cartography can be usefully applied and with a very limited and unimaginative view of the discipline.
>
> GIS is a technique: cartography is a discipline, and by definition cartography must be more than an element in a Geographic Information System. In the convergence between GIS and Computer Assisted Cartography, ... it is important not to lose sight of the importance of cartographic discipline.
>
> Although cartography is an applied discipline there is a need to develop and maintain an intellectual non-applied core and this is largely lacking in the discipline.

[5] Just as someone once observed — I suppose it was Novalis — that romanticism *was* translation, so one might nowadays say that post-modernism *is quotation*. Indeed, to go back to the opening quotation: *Toutes choses sont dites déjà,...*

> Waters, drawing on the work of Bloom and Roszak and others, draws attention to some of the key issues relating to the debate in geography over GIS, which he calls ... the new stampede towards geography's 'Holy Grail' (Waters, 1989, p. 32). He points out that GIS is decidedly positivist and as a result there is little dialogue with either the humanists or the realist/structuralist paradigms in geography.
>
> Cartographers in their quest for scientifically 'objective' products, should not lose sight of the fact that maps have been and continue to be made for a variety of purposes and can never be truly objective because one of their distinguishing features '... is that they focus attention *selectively* on regions of space, features, objects and themes' (Visualingam, 1989, p. 27). Even as new cartographic products and processes emerge this is unlikely to change. The map is an *abstraction* of reality: it is not reality itself. Nor is communication purely objective: it is a *rhetorical* process. At issue is just how explicit the rhetoric is.

The message is clear enough, and it clearly has its own relevance to hydrology. Geography has its Geographic Information Systems (GIS), and these introduce their own *Grammaire générale* into the rhetoric of the subject. It may be very inadequate, but it has at least arrived. And of course this grammar is necessarily established between man and his programmable machine, in a state of symbiosis. We are then naturally led to pose the problem of whether one or other of the models developed in hydrology could fulfil a similar function to that of the GIS in geography (Abbott, 1991b). A natural candidate is the physics-based, spatially-distributed hydrological model, represented by the SHE. Our experience in ecology, however, suggests that the SHE, even in its fourth-generation metamorphosis (Abbott et al., 1991), will not be sufficient in itself for this purpose.

Ecology, like hydrology, has had its fair share of unifying endeavours. We need only recall the attempts of Margalef (e.g. 1968), using notions of information theory and irreversible thermodynamics, to illustrate such attempts. But let us just compare the models of hydrology and ecology at our current point in time, a time of an historical discontinuity in thought. Let us pose the following question: what is the relation between modelling in hydrology and ecology at this point in time?

We see that we have to reply that it corresponds primarily to almost uni-directional orderings, in that a large part of all the results of hydrology are played over to ecology as a basis for ecological modelling, but the results of ecological modelling are returned to hydrology only in small portions and in restricted and mostly other contexts. The reason for this appears at first sight to be quite obvious: the effect of life on land-phase water circulations can apparently be described simply by definite and well-demonstrated, agronomical parameter sets for such features as interception, root water uptake and transpiration. The effects of life thus appear to be rather easily introduced into hydrological models. When seen from the modelling side, however, there is a more fundamental difference. This has to do with the characterisation of ecology as a study of the play of competing *intelligences*

(Abbott, 1991a). It is this feature which then demarcates ecology from hydrology. On the one hand this appears to place ecology in an even more difficult position than hydrology in respect to its theoretical basis, but on the other hand ecology here has the benefit of an already partially-formulated grammar within the massive and variegated productions of theories of artificial intelligence (AI). This provides, once again, for a partial symbiosis between man and his programmable machine, albeit one that must for the present remain only very partial indeed (Abbott, 1991a, b).[6]

Thus, when seen within its bordering subjects, of geography and ecology, hydrology appears to "fall between two stools" in respect to the availability of its definitive true grammar, which constitutes the true rule-matter of its theory. It has still not found this grammar: it is still waiting.

But now we have to ask: is this really such a bad state of affairs after all? For is it not usually better to live with an unformed grammar than with a fully-formed grammar in which one does not truly believe, in which one does not have total faith? Better to be *prepared* for conversion to the true faith than to rush prematurely into a false one. We only have to look at the examples of present-day politics, education and medicine to assure ourselves that an inappropriate grammar, and so an inauthentic theory, may do more harm than good in many places as it arrives at the level of its dialectic, at the level of its practical implementation. Surely, with such salutary examples in mind, hydrology should not be in too much of a hurry in this respect.

On the other hand, since we are in any case coming increasingly to navigate our lives' ways through a world of models, perhaps we could structure a definitive language of hydrology along the lines of the emerging new *Grammaire générale* of the model *per se*. But then, again, the theory of the model is by no means as yet so mature as to support this kind of added responsibility. No, we had better face the facts: there seems to be no simple way at all to redeem the soul of hydrology in the here and now.

---

[6] In fact, in ecology we are confronted with difficulties of communication that are even more intractable than are those of hydrology, since we have in ecology to merge together, through electronic encapsulation, the meanings of utterances of disciplines which are all but impenetrable, the one from the other, so that, even as things stand, the practitioners in the one understand little or nothing of what is said by the practitioners in the other. We have in effect, and then primarily by the use of programmable machines, to merge facts expressed with disparate rhetorics using different grammars and dialectics. What price then a *Grammaire générale*? [And then, just by way of contrast with this situation, see the Fourth Investigation in Husserl's (1913/1970) *Logical Investigations*.] One apparent possibility is to adapt the existing grammar of computational genetics (starting with Crosby, 1974), but this does not take us very far even into the utterances of zoology, botany and their subordinate biologies. After that we have, at most, only a structuralist paradigm (as in anthropology and some parts of sociology) but this is again, to say the least, incomplete: see, once more, Dreyfus and Rabinow (1989). Indeed, we are driven to ask ourselves whether we are really coming, in all this, any further than (re-establishing) a Leibnitzian monadology, and thus still shooting well short even of the Cartesian *res cogitans*. [See, in this respect, Furth (1972), and the references given there.]

6. Does the future hold out any hope at all?

It may be impossible to do very much in the here and now, but perhaps something could be done with hydrology sometime in the future. In order to glimpse a chink of light in this world of darkness, let us take a somewhat other view of the subject: let us regard hydrology as a dialectic played out between science and technology in the domain of their intersection. We can then discern already, at the level of the model, how the definition of science and technology themselves, transform as they pass through our current discontinuity in thought. For, in the 1970s, and indeed well into the 1980s, the emphasis in modelling was placed primarily upon "making our world more accessible through models", and most commonly in mathematical languages (e.g. Abbott and Basco, 1989). From about the mid-1980s onwards, however, a very marked shift of emphasis occurred, towards "making models more accessible within our world". This shift in emphasis was then associated with a new awareness of such matters as user-interface design, database structuring, various "open" architectures and others which were not "scientific" or "technological" in the to-date accepted ways. We had then already to do with quite other scientific perspectives and technological imperatives, more commonly expressed in iconic and natural languages (see Abbott et al., 1991). For those on the outside of this revolution, the whole thing only appeared as a 'repackaging' of existing knowledge, while for those on the inside it involved a complete rethinking of the process of communication between the sciences and the technologies, with profound consequences for the individual sciences and technologies themselves.[7]

It is within this context that it now becomes possible, in particular, to enunciate the notion of an *Enhanced-Solicitude Software Environment* (ESSE). This builds on the increasingly widespread understanding that, above a certain level of complexity, there are no socially-neutral software environments. Work in areas as various as geographic information systems, banking advice servers and hospital work-scheduling systems have all shown that specific software environments and their configurations have quite specific social consequences (Andersen and Mathiassen, 1987; Flores et al, 1988; Fraser Taylor, 1991; Naur, 1992; Wolk et al 1986; more generally, see Eco, 1977, and especially pp. 6, 7, 58 and 64). Although much of the work carried out in this area has tended to accentuate the negative consequences

---

[7] This whole "knowledge industry" is expected to have a turnover of about three trillion (million million) dollars a year by the end of this decade (Kehoe, 1992). A large part of this will then be taken up by *electronic knowledge encapsulators* (EKEs). It is to these that we must increasingly turn in order to eke out our own knowledge resources, which otherwise become ever less adequate to deal with an ever more knowledge-demanding world. (With an eye to the mass market, however, we might perhaps better speak of a *personal* EKE, or a PEKE, thus referring to one of the most dedicated and most playful of man's best friends. Certainly we shall now witness the playing of a game in which some fabulous fortunes will be made and almost equally large amounts lost — which does seem to be the way of the high-tech game generally. And indeed there is nothing very new about this: as Rosa Luxemburg long ago made clear, one of the principal economic functions of higher technology is that of transferring wealth from one social group to another.)

of this situation, it does also open up more positive perspectives. Among these is the possibility of building software environments that tend to enhance the solicitude of the user towards certain values, and in particular those that have already been selected on a basis of policy. In the case of foreign-aid programmes, these values are commonly those of welfare, and specifically the welfare of the poor, of the environment and of women.

The ESSE concept is concerned with structuring software such that its use necessarily inculcates an awareness of values, and so an enhanced state of solicitude towards the bearers of these values. As communication between user and system is still primarily visual, it is supposed in the conceptual design that the user interface will continue to be primarily graphical. The ESSE concept is therefore being currently developed primarily at the level of the graphical user interface. (However, some multi-media possibilities are being investigated.) It is through the design of this level that the desired values are in the first place transmitted. However, the design of the user interface cannot be separated from the overall design of the system, and so, in particular, from its basic architecture. From the work so far carried out, it appears that a multi-agent architecture with message passing amongst agents realised using a hierarchical blackboard, and possibly a concurrent reflective language, will provide the best environment for applying this approach (See Harmon and Mans, 1988, and also Genesereth and Nilsson, 1988). Some suitable database (including knowledge-base) structures, distributed and integrative frameworks and languages have already been identified. It should perhaps be emphasised that ESSE is not itself a tool, language or environment, but is a concept and, potentially, a collection of methodologies, that can be associated with almost any tool, language or environment.

By these means it should become possible to merge facts which correspond to particular instantiations with facts that have essentially constant, fixed attribute values. This corresponds to merging facts that are temporally true (corresponding to the Leibnitzian *sub specie temporalis*) and facts that are true out of time [and so true — and as well, and even more emphatically, in the Spinozistic sense — *sub specie aeternitatis*: see, further, Kneale (1972)]. Could this, perhaps, provide also a path for hydrology?

References

Abbott, M.B., 1991a. Hydroinformatics, information technology and the aquatic environment. Avebury Technical, Aldershot.

Abbott, M.B., 1991b. Contributions of computational hydraulics to the foundation of a computational hydrology. In: D.S. Bowles and P.E. O'Connell (Editors), Recent Advances in the Modelling of Hydrological Systems. Kluwer, Amsterdam.

Abbott, M.B., 1992. Informal methods and magic realism; the limits of modelling and the limits of angelology: the role of scientifically undecidable problems in technology; a primarily theological perspective. In: Dagstuhl Symp. on the Limits of Modelling with Programmable Machines; Formal Models and Reality, Univ. Saarland, Saarbrucken, Extended Synopsis.

Abbott, M.B. and Basco, D.R., 1989. Computation Fluid Dynamics; An Introduction for Engineers. Longman, London, and Wiley, New York., N.Y.

Abbott, M.B. and Minns, A.W, 1993. Computational Hydraulics: Elements of the Theory of Free Surface Flows. 2nd edition, Avebury Technical, Aldershot (in press).

Abbott, M.B., Bathhurst, J.C., Cunge, J.A., O'Connell, P.E. and Rasmussen, J., 1985. An introduction to the European Hydrologic System — Système Hydrologique Européen (SHE). J. Hydrol., 87: 45–59 and 66–77.

Abbott, M.B., Havnø, K. and Lindberg, S., 1991. The fourth generation of modelling in hydraulics. J. Hydraulic Res. 29(5): 581–600.

Andersen, P.B. and Mathiassen, L., 1987. Systems development and use: a science of truth or a theory of lies. In: G. Bjerknes, P. Ehn and M. Kyng (Editors), Computers and Democracy: a Scandinavian Challenge. Avebury, Aldershot.

Barth, K., 1950/1960. Church Dogmatics III, 3. Transl. G.W. Bromiley and T.F. Torrance. Clark, Edinburgh.

Berliner, B., 1982. Limits of Insurability of Risks. Prentice Hall, New York, N.Y.

Chorin, A.J., 1983. Review of R. Peyret and T.D. Taylor, Computational fluid flow. Bull. Am. Math. Soc., 9(3): 368–372.

Cooke, D.J. and Bez, H.E., 1984. Computer Mathematics. Cambridge University Press, Cambridge.

Crosby, J.L., 1974. Computer Simulation and Genetics. Wiley, New York, N.Y. [8]

Dooge, J.C.I., 1973. Linear theory of hydrologic systems. Tech. Bull. No. 1468, Agric. Res. Serv., U.S. Department of Agriculture, Washington, D.C.

Dreyfus, H.L. and Rabinow, P., 1989. Michael Foucault: Beyond Structuralism and Hermeneutics. University of Chicago Press, Chicago, Ill.

Eco, U., 1977. A Theory of Semiotics. Indiana University, Bloomington, Ind.

Flores, F., Graves, M., Hartfield, B. and Winograd, J., 1988. Computer systems and the design of organisational interaction, ACM Trans. Office Automation Systems, 6: 20.

Foucault, M., 1966, Les mots et les choses, une archéologie des sciences humaines. Gallimard, Paris. The Order of Things, an Archeology of the Human Sciences. Tavistock (1970), Routledge (1991), London.

Fraser Taylor, D.R. (Editor), 1991. Geographic Information Systems: The Micro-Computer and Modern Cartography. Pergamon, London.

Furth, M., 1972. Monadology. In: H.G. Frankfurt (Editor), Leibnitz, a Collection of Critical Essays. Notre Dame University, Notre Dame, Ind.

Genesereth, M.R. and Nilsson, N.J., 1988. Logical Foundations of Artificial Intelligence. Morgan-Kaufman, Palo Alto, Calif.

Guay, L. 1990. A multimedia atlas. Proc. Nat. Atlas Information Services Opportunities Sem., September, Department of Energy, Mines and Resources, Ottawa.

Harley, B., 1989. Deconstructing the map. Cartographia, 26(2): 1–20.

Harmon, B. and Mans, R.V., 1988. Expert Systems: Tools and Applications. Wiley, New York, N.Y.

Heidegger, M., 1927/1962. Sein und Zeit. Niemeyer, Tübingen. Being and Time, transl. J. Macquerrie and E. Robinson. Blackwell, London.

Husserl, E., 1913/1970. Logical investigations. Transl. J.N. Findley. Routledge, London.

Kehoe, L., 1992, Bringing wizardry to book. Financial Times, May 1.

Kierkegaard, S.Aa, 1849/1980. Sygdommen til Døden. Collected Works, 2nd edition, Gyldendal, Copenhagen. The sickness unto death, transl. H.V. Hong and E.H. Hong. Princeton University, Princeton, N.J.

Klemes, V., 1986. Dilettantism in hydrology: tradition or destiny. Water Resour. Res., 22(9): 177–188.

[8] Where, as in a time warp, we "oldies" can return to the womb of the old Ferranti Pegasus and the deeper mysteries of its subsequent "Mercury autocode".

Kneale, M., 1972. Leibniz and Spinoza on activity. In: H.G. Frankfurt (Editor), Leibniz, a Collection of Critical Essays. Notre Dame University, Notre Dame, Ind.

Margalef, R., 1968. Perspectives in Ecological Theory. University of Chicago, Chicago, Ill.

Naur, P., 1992. Computing: a Human Activity. ACM Press/Addison Wesley, New York, N.Y., pp. 582–590.

Ramirez, J.R., 1990. Implementation of cartographic language for universal computer-aided mapping systems. ACSM-ASPRS Annual Convention, 2, Washington D.C., pp. 19–29.

Soil Conservation Service, 1975. Soil taxonomy: a basic system of soil classification for making and interpreting soil surveys. Agricultural Handbook No. 436, U.S. Department of Agriculture, Washington, D.C.

Taylor, D.R.F. (Editor), 1985. Education and Training in Contemporary Cartography. Wiley, New York, N.Y.

Visualingam, M., 1989. Cartography, maps and GIS in perspective. Cartogr. J., 26(1): 26–32.

Waters, N. 1989. Bigbytes, microbytes, tidbytes and nibbles: do you sincerely want to be a GISD analyst? Oper. Geogr., 7(4): 30–36.

Wolk, S.R. and Luddy, W.J.L. Jr., 1986. Legal Aspects of Computer Use. Prentice Hall, New York, N.Y.